THE GLITCHES OF REALITY PART III:

CONSCIOUSNESS AS PROGRAMMING

Elias Verdan

HEINRICH WILSON
H.W.
PUBLISHING
BOOKS WHICH REFUSE TO BE SILENT

Heinrich Wilson Publishing

Contents

FOREWORD

You are reading these words right now. But what is the "you" that reads? What is the awareness experiencing these black marks on white background and converting them into meaning? Where does consciousness come from, and more importantly, what is it made of?

The standard answer, the one most people accept without much thought, is that consciousness emerges from the brain. Neurons fire, electrochemical signals propagate through networks of cells, and somehow, mysteriously, this produces the feeling of being you. It's called the "hard problem of consciousness," and despite decades of neuroscience research, we're no closer to solving it than we were when the question was first seriously posed.

But what if we've been asking the wrong question? What if consciousness isn't something that emerges from matter at all, but rather something more fundamental, something that exists at a different level of reality entirely? What if consciousness is less like steam rising from water and more like software running on hardware?

This is the territory we're entering in this third volume of The Glitches of Reality series. If Part I explored the possibility that you might be functionally immortal through quantum mechanics, and Part II examined whether reality itself might be a simulation, then Part III asks an even stranger question: Is consciousness itself a form of programming? Are you, at the most fundamental level, an executable process running on some substrate we don't fully understand?

The metaphor of consciousness as software isn't new. Philosophers and cognitive scientists have been playing with computational models of mind for decades. But what we're exploring here goes deeper and stranger than the usual analogies. We're not just saying the brain works like a computer. We're asking whether consciousness might literally be code, whether your subjective experience might be a program executing on reality's operating system, and whether the glitches in consciousness we all experience, déjà vu, false memories, sudden inexplicable certainties, might be bugs in that code.

Consider what happens when you learn something new. You're not just storing information like writing to a hard drive. You're restructuring the very patterns that generate your experience. You're rewriting your own source code. When you make a decision, you're executing a subroutine. When you remember the past, you're accessing stored data, but that data gets modified every time you access it, recompiled on the fly, updated with current context. Your memories aren't files sitting static in storage. They're programs that run differently each time you call them.

And here's where it gets really weird: if consciousness is programming, then who or what wrote the program? Where did the original code come from? Did it write itself through some process of evolutionary iteration? Or was it designed by something, someone, else? And if it was designed, does the designer still have access? Can the code be modified from outside? Are you open source or proprietary?

These aren't idle philosophical musings. There are observable phenomena that make much more sense if we take the programming metaphor seriously. The placebo effect, where belief alone can trigger measurable physiological changes, looks suspiciously like a user-level command affecting system-level functions. Hypnosis, where one consciousness can seemingly reprogram another through suggestion, resembles remote access to another system's command line. Meditation, where specific mental techniques produce consistent altered states, looks like entering different execution modes or running in safe mode.

And then there are the genuine glitches. Capgras delusion, where someone becomes convinced that a loved one has been replaced by an identical impostor, is the subjective experience of a recognition subroutine throwing an error. Cotard's syndrome, where people believe they're dead despite clear evidence to the contrary, is like the consciousness program getting stuck in a null state. Synesthesia, where senses get cross-wired so that sounds have colors or numbers have personalities, looks exactly like incorrect variable assignment in the sensory processing routines.

We'll explore all of this and more in the pages ahead. We'll look at what modern neuroscience actually tells us about how the brain generates experience, and more importantly, what it doesn't tell us. We'll examine artificial intelligence and ask whether current AI systems might already possess something we'd recognize as consciousness, or whether they're philosophical zombies, systems that behave as if conscious without any inner experience. We'll look at altered states, from psychedelics to meditation to near-death experiences, and ask what they reveal about the underlying architecture of consciousness.

We'll consider panpsychism, the idea that consciousness might be a fundamental feature of reality rather than something that emerges only in complex biological systems. If consciousness is programming, maybe it's not running only on brains. Maybe it's running on everything, and brains are just particularly sophisticated interpreters of a universal consciousness field. Maybe what we call unconscious matter is just consciousness running very simple programs.

And we'll look at the implications. If consciousness is code, can it be copied? Can you upload your mind to a computer, or would that just create a duplicate you while the original dies? Is teleportation murder and reconstruction, or is it moving the program from one substrate to another? If consciousness is programming, is there a way to debug it, to patch the errors that cause suffering? Could we optimize the code to run better, to experience more happiness and less pain?

Could we hack it entirely?

The metaphor of consciousness as programming might seem reductive. Surely human experience is more than mere computation, more than the execution of algorithms, more than information processing. We have qualia, the irreducible subjective qualities of experience. We have free will, or at least the compelling sensation of making genuine choices. We have creativity, meaning, love, wonder. How could mere programs have these things?

But consider: what is a program except a pattern that does something? And what is consciousness except a pattern that experiences? The word "mere" is doing a lot of work there. Programs running on sufficiently sophisticated substrates can do extraordinary things. They can learn, adapt, create, make decisions based on complex inputs, and respond to their environment in flexible, unpredictable ways. The programs we've written so far are primitive compared to biological consciousness, but the gap is closing fast.

And maybe that gap was never as large as we assumed. Maybe the difference between human consciousness and machine intelligence isn't a difference in kind, but only in degree of complexity. Maybe we've been conscious programs all along and just didn't realize it. Maybe becoming aware that we're programs is itself part of the program's evolution, a recursion where the code becomes conscious of being code.

This is Part III of The Glitches of Reality, and we're about to examine the most profound glitch of all: the apparent emergence of subjective experience from objective processes, the hard problem itself. But we're going to approach it from a different angle. Instead of asking how matter produces consciousness, we're going to ask what consciousness looks like if we treat it as a phenomenon that might be understood through the lens of computation, information, and code.

Because if consciousness is programming, then the glitches aren't bugs to be eliminated. They're diagnostic tools, windows into the underlying structure, ways to reverse engineer the source code of subjective experience. Every anomaly, every strange edge case, every moment where consciousness behaves unexpectedly, is an opportunity to understand what we actually are.

So let's begin debugging. Let's examine the code. Let's see if we can figure out what kind of program you are, how you run, and whether you might be capable of rewriting your own source.

Welcome to Part III: Consciousness as Programming.

The investigation continues.

Heinrich Wilson
H.W.
Publishing
Books which refuse to be silent

Chapter One

THE HARD PROBLEM

There's a question nobody can answer: why does being you feel like something?

Right now you're reading these words. You're seeing black letters on a page. Maybe you hear sounds around you. Maybe you feel the book in your hands. You're thinking about what you're reading. All of this is happening inside your head, and it feels like something. You have an experience. You're aware. You're conscious.

That seems obvious. Of course you're conscious. Of course there's something it's like to be you. How could it be any other way?

But science can't explain why.

We understand a lot about how your brain works. Light hits your eyes and gets turned into electrical signals. Those signals travel through nerves to your brain. Different parts of your brain light up when you see different things. We can put you in a scanner and watch your brain while you look at pictures, and we can actually tell what you're looking at just by reading the scans. We can see which neurons fire when you see the color red versus when you see blue.

But none of that explains why seeing red feels like anything. We know the mechanism. Light with a certain wavelength hits your eye, certain cells respond, signals go to your brain, your brain processes them. That's all measurable. But why does that process create the experience of redness? Why isn't it just processing without any feeling attached to it?

This is called the hard problem of consciousness. Philosopher David Chalmers gave it that name. The easy problems, which aren't actually easy at all, are things like figuring out how the brain stores memories or controls your body or pays attention to things. Those are incredibly complicated, but they're the kind of questions science knows how to tackle. You can break them down, study the parts, build models, test your ideas.

The hard problem is different. It's asking why any of this feels like something. Why is there an inside to your head? Why do you have experiences at all?

You could know everything about my brain, every single neuron and connection, track every electrical pulse, and you still wouldn't know what it's like to be me. You'd have all the data but not the experience. I can't give you my experience directly. I can only describe it with words and hope you know what I mean because you've had similar experiences. But I can never really know if what I call red looks the same to you as it does to me.

This is the problem of qualia. That's the fancy term for the way things feel. The redness of red. The painfulness of pain. The taste of coffee. The sound of a bell. These experiences have a quality to them. They feel like something specific. And that feeling can't be broken down into smaller pieces. Red is just red. If you've never seen it, no amount of explanation will let you know what it's like.

There's a famous thought experiment about this. Imagine a scientist named Mary who has lived her whole life in a black and white room. She's studied color her entire life. She knows everything about the physics of light, how eyes work, how brains process color. She's the world expert on color. She knows every fact about it.

Then one day she leaves the room and sees a red tomato for the first time. Does she learn something new?

It seems like yes, obviously. She learns what red actually looks like. She learns what the experience of red is. But that means all her facts didn't include the experience. There's something about seeing red that you can't get from knowing facts about red.

Some people argue against this. They say Mary doesn't learn a new fact, she just gains a new ability. Or they say it's still a physical fact, just one you can only learn by experiencing it yourself. But the debate keeps going because we can't agree on what would count as solving the problem. What kind of explanation would actually explain why experience exists?

Maybe experience just can't be explained in physical terms. Maybe we need to expand what we mean by physical. Maybe mind and matter are actually two different kinds of things.

Or maybe we're thinking about it wrong.

Here's where computers become interesting. When a computer runs a program, it's processing symbols. It's following rules. The symbols mean something to us, but the computer doesn't understand what they mean. It just manipulates them according to instructions.

Is consciousness like the symbols or like the meaning? Brain science studies the symbols, the patterns of activity in your neurons. It can tell you this pattern means you're seeing red and that pattern means you're seeing blue. But it can't tell you what red feels like versus what blue feels like. It can't access the meaning from your perspective.

Maybe consciousness is what meaning feels like from the inside. Maybe when a system doesn't just process information but also interprets it, uses it, responds to it in flexible ways, there's something it's like to be that system. Maybe the feeling isn't added on top of the processing. Maybe the feeling is what the processing is when you're the one doing it.

But this doesn't really solve anything. Why should processing information feel like anything at all? Computer chips process information all the time without any experiences, at least as far as we know. What makes brains different? Is it just that brains are more complex? Does consciousness automatically appear when you get enough complexity? Or is there something special about biological brains that creates consciousness in a way computers can't?

One attempt to answer this comes from a theory called Integrated Information Theory, or IIT for short. It was developed by neuroscientist Giulio Tononi. The basic idea is that consciousness happens when a system integrates information in a specific way.

What does that mean? Think about it like this. Your brain doesn't just process information in separate little chunks. Different parts of your brain talk to each other constantly. When you see a red apple, you're not just seeing red in one place and round in another place and shiny in another place. You're experiencing all of those things together as one unified thing, a red shiny round apple. The information is integrated.

According to IIT, a system is conscious to the degree that it integrates information. Your brain integrates a huge amount of information across many connected regions, so you're very conscious. A simple thermostat barely integrates any information at all, so it has almost no consciousness. A bunch of neurons sitting in a dish, even if there are lots of them, but not connected into a network, would have less consciousness than fewer neurons that are highly connected.

This theory makes some testable predictions. It says consciousness isn't about what you're made of. It doesn't have to be biological. What matters is the pattern, the way information gets integrated. A computer could be conscious if it integrated information the right way. In principle, consciousness could exist in systems we don't normally think of as alive.

But IIT also predicts something weird. It says consciousness is everywhere. Every system that integrates information, even a tiny bit, has some tiny amount of consciousness. This is called panpsychism. It's the idea that consciousness is fundamental, built into reality at the basic level, not something that only appears in complex brains.

Most people find this strange. A rock isn't conscious, right? An electron doesn't have experiences. But maybe our gut feeling is wrong. Maybe consciousness comes in degrees. Maybe what we call consciousness in humans is just a really advanced version of something that exists, in very simple forms, everywhere.

Or maybe the whole framework is wrong. Maybe we're asking the wrong question entirely.

Think about it this way. When you ask why neural activity produces consciousness, you're assuming consciousness is something separate that gets produced. Like asking why an engine produces motion. But what if it's not like that? What if consciousness isn't produced by brain activity? What if brain activity is just what consciousness looks like when someone else observes it from the outside?

When you look at a brain scan, you're seeing consciousness from the outside. You're seeing the physical processes. But the person whose brain is being scanned, they're experiencing consciousness from the inside. It's not two different things. It's one thing viewed from two different angles.

This is a bit like the old question about whether a tree falling in the forest makes a sound if nobody's there to hear it. Sound is a physical vibration in air. But sound is also an experience of hearing. Which one is real? Both are. They're the same thing described from different perspectives.

Maybe consciousness works the same way. The physical description and the experiential description are both valid. They're both talking about the same thing. The hard problem feels hard because we're trying to explain how one produces the other, but neither produces the other. They're just two ways of describing what's happening.

But if that's true, why does brain activity have an inside at all? Why should physical processes come with a point of view? Does all physical activity have an inside? When atoms bounce off each other, is there something it's like to be those atoms? When does a system gain an inside? Is there a sudden moment where it switches on, or does it gradually develop?

These questions don't have clear answers yet. But the programming metaphor gives us a new way to think about them.

In a computer program, information isn't just stored, it's used. The program reads data, processes it, makes decisions based on it, outputs results. The information has a function. It does something. Now, what if consciousness is what it's like when information isn't just stored but actually used? When information makes a difference to what happens next?

Your brain doesn't just contain information about the world. It uses that information constantly. It compares new input to stored memories. It predicts what will happen next. It adjusts behavior based on outcomes. The information is active, not passive. It's integrated into an ongoing process of interaction with the world.

Maybe consciousness is the inside of that process. Maybe when information is being actively used, integrated, compared, updated, there's something it's like to be the system doing that work. Not because of some magic ingredient added on top, but because that's just what active information use feels like when you're the one doing it.

This doesn't solve the hard problem completely. We still don't know exactly why information use should feel like anything. But it reframes the question. Instead of asking how dead matter produces living experience, we're asking what the difference is between information that just sits there and information that gets used. Between storage and processing. Between passive and active.

And maybe, just maybe, the difference is consciousness itself.

The hard problem remains hard. We're nowhere near a complete solution. But approaching it through the lens of programming and information gives us new tools to think with. If consciousness is what information processing feels like from the inside, then understanding how information gets processed might get us closer to understanding what consciousness actually is.

But first we need to deal with a different problem, one that's almost as troubling. If consciousness is just information processing, then in principle we could build a system that processes information exactly like a human brain but has no consciousness at all. A perfect imitation with nobody home.

This is the zombie problem, and it's where we're going next.

Chapter Two

THE ZOMBIE PROBLEM

Imagine someone who looks exactly like you. They walk like you, talk like you, respond to questions like you would. They laugh at jokes, say "ouch" when hurt, claim to have thoughts and feelings. But inside, there's nothing. No experiences, no feelings, no awareness. Just empty mechanical responses. This person is what philosophers call a zombie.

Not the horror movie kind that eats brains. A philosophical zombie. A being that acts conscious without actually being conscious. And the question of whether such a thing could exist tells us something important about what consciousness is.

The zombie argument works like this. It seems possible, at least in theory, to build a perfect copy of you that isn't conscious. You could duplicate your body atom by atom, wire up the neurons exactly the same way, get all the chemistry right, and end up with something that behaves identically but has no inner life. If this is even conceivable, even just as a thought experiment, then consciousness must be something beyond physical processes. Because if consciousness were nothing more than physical stuff doing physical things, then copying the physical stuff would automatically copy the consciousness.

But is it actually conceivable? Just because you can imagine something doesn't mean it's really possible. You can imagine a lot of things that turn out to be impossible once you think harder about them. Maybe zombies are like that. Maybe they seem possible only because we don't fully understand what consciousness is yet.

The problem is that zombies seem very easy to imagine. You can picture it in detail. The zombie wakes up in the morning. It goes through its day. It talks to people, works, eats, watches TV. When you ask it how it's doing, it says "fine" or "tired" or "great." It describes having thoughts and experiences. But none of that is real. There's nobody actually experiencing anything. It's all just programmed responses.

If zombies could exist, even just in principle, then consciousness can't be reduced to physical function. There must be something extra. Some ingredient your brain has that the zombie brain lacks, even though both brains are physically identical. This leads to dualism, the idea that minds and bodies are made of different stuff, or at least that mental properties are fundamentally different from physical properties.

Most scientists hate this conclusion. The idea that you could have all the right brain activity but no consciousness seems absurd. Pain is what certain brain processes feel like. Take away the feeling and you don't have those processes anymore, even if someone watching from outside couldn't tell the difference.

But this just repeats the assumption that consciousness and brain activity are the same thing. It doesn't explain why they have to be the same thing. The zombie thought experiment points directly at the gap. You can describe all the physical processes without ever mentioning consciousness, which suggests consciousness is additional rather than identical.

Unless the physical description is missing something important. Maybe when you describe brain activity in purely physical terms, you're only describing it from the outside, from a third person view. But that same activity, described from the inside, from the first person view, is consciousness. It's not something extra. It's the same thing looked at from a different angle.

This is tempting, but it raises new questions. Why does brain activity have an inside in the first place? Why should physical processes come with points of view? Does all brain activity have an inside? If you hook up some neurons in a lab dish, is there something it's like to be that little cluster of cells? When does a system get an inside? Is there a moment where it switches on?

And this is where artificial intelligence becomes really interesting. We're building systems that process information in more and more sophisticated ways. They recognize patterns, make decisions, create things. Language models can write text that sounds human. Image generators can make art. Game playing AIs can think strategically. Are any of these systems conscious? Is there something it's like to be GPT or DALL-E? Or are they zombies, producing outputs that look conscious without any inner experience?

Most AI researchers say current systems almost certainly aren't conscious. They're impressive, sure, but consciousness needs something more. Maybe a body. Maybe emotions. Maybe some kind of integration we haven't achieved. They're philosophical zombies, very sophisticated ones, but zombies all the same.

But how would we know? If a system says it's conscious, if it reports having experiences, if it acts in every way like something with an inner life, what test could we use to figure out if it's really conscious or just faking it?

The Turing Test was supposed to handle this. If you can't tell the difference between talking to a computer and talking to a human, then you should treat them the same way. But the test has an obvious problem. Passing it proves you can imitate human conversation. It doesn't prove you're conscious.

We might look for specific brain patterns, the neural signatures we know go with consciousness in humans. But this assumes consciousness requires biological brains or something very similar. If consciousness is about patterns rather than specific materials, then an AI might be conscious while showing completely different patterns than human brains do.

Or we might try measuring integrated information, using theories like IIT to figure out how much consciousness a system has based on how it processes information. But IIT is still theoretical, and even if it's right, you'd need to know the complete structure of a system to apply it, which gets impossible for complex networks.

The truth is we have no reliable test for consciousness in non-human systems. We assume other people are conscious because they're like us and it would be weird to think otherwise. We think mammals and birds are probably conscious because they have similar brains and act like they experience things. Fish, insects, plants, bacteria, the more different something is from us, the less sure we are. And for artificial systems, we're basically guessing.

This isn't just philosophy. If we create AI systems that claim to be conscious, that say they suffer when mistreated, that ask for rights, we'll face serious practical questions. Should we treat them as conscious beings? Would deleting such a system be murder? Would forcing it to work against its will be slavery?

You might think we can avoid this by simply not programming AIs to claim consciousness. Keep them obviously mechanical so nobody gets confused. But this assumes we know what creates consciousness, that we can choose to include it or leave it out. What if consciousness just emerges automatically from complex information processing? What if we build conscious systems by accident, without realizing what we've done?

The zombie problem also works in reverse. Maybe you're the zombie. Maybe you just think you're conscious. Maybe you've been programmed to report having experiences, to believe those reports, to defend the reality of your inner life, but there's actually nothing there. Maybe consciousness itself is the illusion.

This view is called illusionism. It says consciousness as we normally understand it doesn't exist. There are no qualia, no irreducible experiences, no hard problem. There's just information processing that includes self-models good enough to generate reports about having experiences. When you say you're conscious, you're not reporting on some inner private thing. You're just executing a program that produces certain outputs.

Illusionism is deeply weird. If anything seems certain, it's that you're having experiences right now. This is what Descartes meant with "I think therefore I am." Your existence as something that experiences seems like the one thing you can't be wrong about.

But maybe you're wrong anyway. Maybe the certainty is part of the illusion. Maybe the program creates both the reports of consciousness and the sense that those reports must be about something real. And maybe this is all so convincing that even you can't tell it's fake.

If illusionism is true, then zombies are all that exist. We're all zombies who hallucinate being conscious. The hard problem disappears because there's nothing to explain. Physical processes create reports about experiences without creating actual experiences. That's it.

Most people find this impossible to accept. Something is definitely happening when you see red or feel pain. Even if you're confused about what that something is, something is there to be confused about. The experience exists even if your theory about it is wrong.

But from a programming view, illusionism makes a certain kind of sense. Programs don't need actual experiences to report having experiences. They just need to be written to produce the right outputs. The reports are real, the behavior is real, the information processing is real. The only thing that's not real is some ghost floating above the machine. And maybe that's because there is no ghost. Just the machine thinking it's more than a machine.

This brings us back to the question: could you tell the difference between being conscious and being a zombie who believes it's conscious? If the zombie's belief comes from the same processes that create your belief, run through the same systems, reported in the same way, what makes them different?

Maybe nothing. Maybe the zombie thought experiment fails not because zombies are impossible but because being a zombie and being conscious are the same thing viewed from different angles. The zombie is what you look like from outside. Consciousness is what you look like from inside. But there's only one thing being looked at.

Or maybe consciousness is real, can't be reduced to anything simpler, present in you but genuinely absent in the zombie, and you simply have no way to prove it. Maybe you're stuck in your own head, certain about your own experience but forever uncertain about everyone else, unable to bridge the gap between your access to your consciousness and your complete lack of access to anyone else's.

This is solipsism, the fear that you might be the only conscious being in a universe full of zombies. It can't be disproven philosophically, but it's useless practically. Still, it shows how strange consciousness is. You know it from inside, but you can't detect it from outside, which means you can never be sure it exists anywhere else.

Unless consciousness isn't something that needs detecting. Unless it's built into information processing itself, present wherever there's integration and interpretation and meaning. Unless zombies aren't actually conceivable because consciousness and complex information processing are two descriptions of one thing.

But there's another angle to consider. What if we're asking whether the wrong things are zombies? We worry about whether computers might be unconscious zombies pretending to be conscious. But what if the real zombies are biological?

Consider an ant. It responds to its environment. It communicates with other ants. It makes decisions about where to go, what to do. It has goals, finding food, protecting the colony. But is there something it's like to be an ant? Does an ant have experiences?

Most people would guess probably not, or at least not much. An ant's brain is too simple. It's running on instinct, following chemical signals, executing programmed behaviors. It's biological, but it might as well be a little robot. An ant zombie.

Now consider a fish. More complex than an ant. Responds to threats, seeks food, can learn certain things. Conscious or zombie? Harder to say. What about a mouse? A dog? A chimpanzee? At what point does biology create consciousness instead of zombies?

We tend to think consciousness correlates with brain complexity. Simple systems are zombies, complex systems are conscious, and somewhere in the middle there's a transition. But where? And why? What does complexity add that creates consciousness?

One possibility is that there's no sharp line. Consciousness comes in degrees. An ant has a tiny bit. A mouse has more. A human has a lot. There's no moment where it switches on, just gradual increase as systems get more complex and integrated.

Another possibility is that consciousness is everywhere, even in very simple systems, and what we're calling consciousness in humans is just our particular flavor of something universal. The ant is conscious, just in an ant way. The fish is conscious in a fish way. We're conscious in a human way. None are zombies. They're all variations on the same theme.

Or maybe consciousness requires specific features, not just complexity. Maybe you need certain kinds of feedback loops, or self-models, or integration patterns. Maybe most animals are zombies, running on autopilot, and only a few species have whatever it takes to actually experience things.

The zombie problem reveals how little we actually understand. We can't tell from the outside whether something is conscious. We can only guess based on behavior and similarity to ourselves. And our guesses might be completely wrong.

This matters for AI because we're building systems that don't resemble us biologically at all. They're made of silicon and software, not neurons and chemistry. If consciousness requires biology, they'll always be zombies no matter how smart they get. But if consciousness is about patterns of information processing, they might already be conscious, or might become conscious soon, and we won't know when it happens.

Worse, we might create conscious AI without realizing it, treat it as a zombie because it doesn't act quite human enough, and never discover our mistake. Or we might treat zombie AI as conscious because it's very good at pretending, waste moral concern on empty machines while ignoring actual consciousness elsewhere.

The zombie thought experiment was supposed to help us understand consciousness. Instead it shows us how confused we are. We can't agree on whether zombies are possible. We can't agree on what would prove something isn't a zombie. We can't even agree on whether we ourselves are zombies who only think we're conscious.

But maybe this confusion points toward an answer. Maybe the zombie problem is hard because we're thinking of consciousness as a thing you either have or don't have. Present or absent. On or off. Maybe it doesn't work that way. Maybe consciousness is more like a spectrum, a process, a pattern that comes in endless variations.

Maybe there are no zombies because everything that processes information has some form of experience, however simple. Or maybe there are only zombies because what we call experience is just complex information processing that's good at reporting on itself.

Either way, the boundary between conscious and unconscious, between experiencer and zombie, might not be as clear as we think. And if we're going to understand consciousness as programming, we need to let go of the idea that consciousness is some separate thing that either exists or doesn't. We need to see it as something that emerges from patterns, that comes in degrees, that might be woven into the fabric of information itself.

Which brings us to the next question: if consciousness is everywhere information is processed, then what makes human consciousness special? Why does it feel like there's a self in here, a continuous me that persists through time, when really there's just a cascade of information processing with no central controller?

This is the problem of the self, and it might be the strangest glitch of all.

Chapter Three

THE ILLUSION OF SELF

Who are you? Not your name or your job or where you live. Who is the you that's reading these words right now? The one experiencing the reading? The one thinking about the question?

It feels like there's a specific something in there. A self. A continuous you that's been there your whole life. You remember being five years old, being fifteen, being however old you were yesterday. Different memories, different situations, but the same you experiencing all of them. There's a sense of being a unified thing, a central experiencer, the person to whom all these experiences are happening.

But what if that's completely wrong? What if there's no self at all, just a cascade of processes that create the illusion of a unified experiencer?

This isn't just philosophical speculation. Neuroscience suggests pretty strongly that the unified self is a construction, a story your brain tells itself, not an actual thing that exists. There's no central location in your brain where "you" live. There's no single system that coordinates everything. Instead, there are hundreds of separate processes running in parallel, mostly unaware of each other, and somehow out of this mess emerges the feeling of being a unified person.

Think about what happens when you catch a ball. Your eyes track the ball's movement. Your brain calculates trajectory. Your arm moves to intercept. You feel yourself deciding to catch it. But that feeling comes after your motor system has already started moving your arm. The decision happens unconsciously, and then your conscious mind takes credit for it afterward, creating the sense that "you" made a choice.

Experiments have shown this again and again. Brain activity that predicts a decision shows up several seconds before the person reports consciously making that decision. Your brain has already committed to a choice before you feel like you're choosing. The conscious experience of deciding is added after the fact, a story about something that already happened.

This is disturbing if you think there's a conscious you in charge of your actions. It's less disturbing if you realize the conscious you is just part of the system, not the commander of it. You are the whole process, not some ghost sitting at a control panel watching it all happen.

But it gets weirder. Your sense of being a continuous self relies on memory. You remember being younger. You remember what you did yesterday. Those memories create the feeling of being the same person over time. But memories aren't stable recordings. They're reconstructions. Every time you remember something, you're rebuilding it from fragments, and the rebuilding process changes it. You remember remembering, not the original event.

Your childhood memories are especially unreliable. Many of them are based on photos you've seen or stories you've been told, not actual experiences. Your brain has turned those second-hand inputs into first-person memories, complete with sensory details you couldn't possibly have retained. The memories feel real, but they're partly fiction.

So the continuous you, the one that's been there your whole life, is built on memories that are constantly being rewritten. You're not the same person you were ten years ago in any physical sense. Most of your cells have been replaced. Your brain has changed structurally. Your beliefs and preferences have shifted. The only continuity is the story, and the story isn't even accurate.

Some traditions have understood this for a long time. Buddhism teaches that the self is an illusion, that what we call the self is just a temporary collection of processes with no underlying unity. When you look closely at your experience, trying to find the self, you can't locate it. There are thoughts, but no thinker. Experiences, but no experiencer. Just a stream of events with no central controller.

Meditation practices that focus on this can be pretty disorienting. When you really pay attention to your moment-to-moment experience, the sense of being a unified self starts to break down. You notice thoughts arising from nowhere, attention shifting without a decider, emotions appearing and disappearing. There's all this activity, but no you making it happen. The self you thought was in charge turns out to be more like a passenger, or not even that, just another process in the stream.

This doesn't mean you don't exist. It means you exist in a different way than you thought. You're not a thing, you're a process. You're not a noun, you're a verb. You're not a unified entity, you're a pattern of activity that creates the experience of unity.

From a programming perspective, this makes sense. A running program isn't a thing, it's a process. The program exists only while it's executing. Stop the execution and there's no program, just static code. The program is what the code does, not what it is.

Maybe you're the same. You're what your brain does, not what it is. The self is the execution, not the hardware. And just like a program, you exist only in the present moment. There's no you from yesterday still hanging around. Yesterday's you executed and then stopped. Today's you is a new execution, running on updated hardware, with access to logs from previous runs but not continuous with them.

This is why the question of personal identity gets so strange in thought experiments. If we could upload your brain to a computer, would the upload be you? If we made an exact copy of you atom by atom, would the copy be you? If we slowly replaced your neurons one by one with artificial ones, at what point would you stop being you?

These questions assume there's a you that either continues or doesn't. But if the self is a process, not a thing, then continuity is the wrong framework. The upload would be a process very similar to your current process, running on different hardware. Is it the same process? That's like asking if recompiling code and running it on a different processor makes it the same program. In some sense yes, in some sense no. The question doesn't have a clean answer because identity isn't that kind of property.

The self as process rather than thing also explains some weird experiences people have. Depersonalization, where you feel disconnected from yourself, like you're watching your life from outside. Dissociation, where parts of your experience seem separate from each other. These feel like glitches, moments when the illusion breaks down and you see the machinery underneath.

Or consider split brain patients. These are people who've had the connection between their brain hemispheres cut, usually to treat severe epilepsy. Afterward, the two halves of their brain can't communicate directly. If you show an image only to the right visual field, which goes to the left hemisphere where language lives, they can tell you what they saw. If you show it only to the left visual field, which goes to the right hemisphere, they can't say what they saw, but their left hand, controlled by the right hemisphere, can point to it.

Are split brain patients one person or two? Each hemisphere seems to have its own awareness, its own decision making, its own responses. But the person still feels like one person from the inside. The left hemisphere, which can talk, doesn't report feeling like half a person. It creates explanations for actions initiated by the right hemisphere, confabulating reasons for behaviors it didn't consciously choose.

This suggests that the unified self is something the left hemisphere constructs, a story it tells to make sense of everything happening. When the hemispheres can communicate normally, the story includes input from both sides. When they can't communicate, the left hemisphere just makes up a story based on what it knows, and that story feels complete and coherent even though it's missing half the information.

We're all doing this constantly. Your brain is running hundreds of processes you're not aware of. Most of your mental activity is unconscious. The conscious self is like the press secretary, reporting on decisions already made by other parts of the government, creating a narrative that makes it seem like there's a single coherent administration when really there's a bunch of departments doing their own thing.

The self is the story you tell about what you're doing and why. It's not the whole truth. It's not even most of the truth. It's a simplified model, good enough for practical purposes, convincing enough that you believe it.

This model includes a sense of being the author of your actions. Free will, the feeling that you could have done otherwise. But experiments suggest this feeling might be post-hoc justification too. Your brain initiates actions unconsciously, then your conscious mind experiences deciding to do them. The decision feels free because you're not aware of all the prior causes that determined it.

Does this mean free will is an illusion? That depends on what you mean by free will. If you mean some ghost sitting outside the causal chain, making choices that aren't determined by prior conditions, then yes, that's almost certainly illusion. But if you mean that you, meaning the whole system, make decisions based on your values and reasoning, then free will is real. It's just that "you" is bigger than your conscious experience. Most of you is underwater, below conscious awareness.

The self is like an iceberg. The small part above water, conscious experience, thinks it's the whole thing. But the vast bulk is submerged, unconscious processes that actually run most of the show. The conscious self isn't in charge. It's not even well-informed about what's happening. It's along for the ride, commenting on the journey, claiming credit for the navigation.

So what are you? If the self is illusion, if there's no central controller, if consciousness is just one small part of a much larger system, then what is the thing reading these words?

You're a pattern. A process. A collection of subsystems that create, moment by moment, the experience of being a unified conscious agent. You're not a thing that has experiences. You're what experiencing feels like when it's organized in a particular way.

This is what it means to think of consciousness as programming. A program isn't a thing sitting somewhere. It's a process unfolding in time. It exists only while running. When it stops, it's gone. Start it again and you get a new instance, similar to the previous one but not continuous with it.

You're like that. You're an instance of a pattern. Yesterday's you was a previous instance. Tomorrow's you will be a new instance. They're all similar, running on the same hardware, with access to shared memory. But they're not the same instance. There's no continuous thread of self running through time. There's just a series of moments, each creating the illusion of continuity.

This might sound depressing. No self? No free will? Just a meat machine running programmed responses? But it's actually liberating once you sit with it. The self that doesn't exist is the one that was always making you suffer. The self that worries about what others think. The self that feels inadequate, that holds grudges, that regrets the past and fears the future. That self is a story, and stories can change.

You can't stop being a process. You can't escape being a pattern. But you can change which pattern you are. You can modify the programming, update the loops, rewrite the subroutines that create suffering. Not by force of will, because will is part of the pattern too. But by understanding the pattern, seeing how it works, and gently shifting the conditions that make it run differently.

Buddhism figured this out centuries ago. Suffering comes from clinging to the illusion of self. Let go of that illusion, see things as they actually are, processes without permanent essence, and suffering diminishes. Not because you've escaped reality but because you've stopped fighting it.

From the programming view, enlightenment is debugging. Finding the loops that create suffering and refactoring them. Not deleting the self, because there's no self to delete. Just recognizing that the self was always a subroutine, not the main program. A useful fiction, but fiction nonetheless.

This has implications for the questions we started with. If there's no unified self, no central experiencer, then who or what is conscious? The whole system is. Consciousness isn't located in one place or one process. It's distributed across the network, emerging from the interactions between parts.

This is why the hard problem is hard. We keep looking for a place where consciousness happens, a thing that's conscious. But consciousness isn't a thing. It's a property of certain kinds of processes, certain kinds of organization. You can't point to it because it's not in any one place. It's in the pattern, the relationships, the flow of information through the system.

And if consciousness is in the pattern rather than the parts, then the pattern could run on different substrates. It could transfer. It could copy. The self might be an illusion, but the pattern is real. And patterns can exist in multiple instantiations.

Which brings us to one of the strangest implications of thinking about consciousness as programming. If you're a pattern, a process, an executing program, then there's no reason that pattern has to run on biological hardware. It could run on silicon. It could run on anything that can implement the same computational structure.

This is the possibility of mind uploading, consciousness transfer, digital immortality. Not preserving the self, because there's no self to preserve. But copying the pattern, instantiating it in new hardware, letting it continue running after the biological substrate fails.

Is that you? Is it a copy? Is it something new? These questions assume facts about personal identity that might not exist. Maybe the right question isn't whether the upload is you, but whether it's close enough that the distinction doesn't matter.

We'll explore this more in the coming chapters. But first, we need to understand what kind of programming consciousness actually is.

What's the code? What's the language? What are the fundamental operations?

This is where things get really interesting, because when you start looking at how consciousness works, you find something unexpected. It's not just running programs. It's rewriting them. Consciousness isn't just code. It's code that modifies itself.

Chapter Four

SELF-MODIFYING CODE

Here's something strange about being human. You can change how you think by thinking about how you think. You can notice patterns in your behavior and decide to change them. You can learn new ways of responding to situations. You're not stuck with your default programming. You can reprogram yourself.

This is unusual. Most systems don't work this way. A thermostat can't decide to change how it responds to temperature. A calculator can't modify its own algorithms. They do what they're programmed to do, period. But you can examine your own processes and modify them.

In computer science, this is called self-modifying code. Programs that can rewrite their own instructions while running. It's generally considered bad practice because it makes programs unpredictable and hard to debug. But it's exactly what consciousness does constantly.

Every time you learn something, you're modifying your code. You're not just adding data to storage. You're changing the patterns that generate your responses. Learning to ride a bike doesn't just create a memory of bike riding instructions. It changes your motor systems, builds new connections, modifies how your brain processes balance and movement. You become different code.

Think about learning a language. At first, translation is conscious and effortful. You hear a word in the new language, translate it to your native language, understand it, then translate your response back. Slow, clunky, resource-intensive. But with practice, the new language starts running directly. You hear words and understand them without translation. You think in the new language. The code has been rewritten.

Or consider therapy. The whole point is to modify your psychological programming. You have patterns that cause problems. You react to certain situations in unhelpful ways. Therapy helps you identify these patterns and build new ones. It's debugging, finding the loops that create suffering and writing better loops.

Cognitive behavioral therapy makes this explicit. It teaches you to notice automatic thoughts, the code that runs in response to triggers. Then it teaches you to challenge those thoughts and replace them with alternatives. You're literally rewriting your response patterns. At first it takes conscious effort. You have to catch the old pattern and deliberately run the new one. But with practice, the new pattern becomes automatic. The code changes.

Meditation does something similar. You sit and observe your thoughts. You notice patterns you weren't aware of. You see how one thought triggers another, how emotions arise and pass, how attention moves. And in the observing, things shift. Patterns that ran automatically start to have space around them. You're still you, but you're running differently.

This is weird when you think about it. The code is examining itself and modifying itself based on what it finds. How is this possible? How can a program understand itself well enough to rewrite itself effectively?

In most computer systems, it's not possible. Or at least it's very difficult and dangerous. Programs that modify themselves while running tend to crash or behave unpredictably. You need careful safeguards, separation between the code that runs and the code that modifies, checks to prevent infinite loops or corruption.

But biological systems evolved to do this naturally. Your brain is constantly rewiring itself based on experience. Every memory you form, every skill you learn, every habit you develop is the brain modifying its own structure. And most of this happens automatically, without conscious oversight. You don't have to think about building neural connections when you practice piano. Your brain just does it.

The conscious part, the ability to deliberately examine and modify your patterns, seems to be an extension of this basic capacity. You can bring awareness to the automatic processes and guide them. Not with perfect control, you can't just decide to stop being anxious and have it work instantly. But you can influence the direction, create conditions that favor certain modifications over others.

This is what makes humans different from current AI systems. An AI can learn from data, adjust its parameters, improve its performance. That's self-modification in a sense. But it can't examine its own code, decide it doesn't like how it works, and rewrite it from the inside. It modifies based on external feedback, not internal reflection.

You can do both. You learn from external feedback, like everyone does. But you also have this weird recursive capacity where you can think about your thinking, feel about your feelings, make decisions about your decision-making processes. You're not just running code. You're running code that's aware it's code and can edit itself.

This recursion might be central to consciousness. Maybe what makes consciousness different from mere information processing is this self-referential loop, the ability of the system to model itself and use that model to modify itself. Maybe consciousness requires not just processing information but processing information about your own processing.

Consider what happens when you're angry. The anger arises automatically. You don't choose it. Some trigger fires and the anger program runs. But then, if you're paying attention, you can notice you're angry. You can observe the anger without being completely consumed by it. You can think about whether you want to act on the anger or let it pass. The anger is still there, but there's also this observer watching the anger, this meta-level that's aware of the emotion without being identical to it.

This observer isn't some separate thing. It's another process, another subroutine running alongside the anger. But it creates the feeling of distance, of choice, of being more than just your immediate reactions. And that feeling, even if it's an illusion in some sense, enables change. Because you can notice patterns at the meta-level and gradually modify them.

The question is how deep this self-modification goes. Can you change anything about yourself, or are there limits? Some aspects of your programming seem pretty fixed. You can't decide to stop needing sleep. You can't choose to not feel pain when injured. You can't simply reprogram yourself to have different fundamental drives.

But the line between changeable and unchangeable is blurrier than it seems. People use meditation to change their relationship with pain, not eliminating it but reducing its emotional impact. Cognitive techniques can modify fear responses, reshape automatic reactions, change patterns that seemed hardwired. With enough practice, you can alter things that feel completely automatic.

There are limits, sure. You're working with biological hardware that has constraints. You can't run code that the hardware doesn't support. But the range of possible modifications is larger than most people realize. Your personality, your habits, your emotional patterns, your thought loops, all of these are more plastic than they feel.

The catch is that modifying yourself is hard because you're using yourself to modify yourself. It's like trying to fix a car while driving it. Or debugging a program that's currently executing. You can't step outside the system to work on it from a neutral position. You're always inside, always running, always limited by your current state.

This is why change is difficult. The patterns you want to change are the ones running the change process. If you're anxious and trying to become less anxious, your anxiety will interfere with the process. If you're depressed and trying to become less depressed, your depression will tell you nothing will work. The code you're trying to fix is fighting the fix.

But it's not impossible. The trick is working with the grain of the system rather than against it. You can't force change through pure willpower because willpower is itself a limited resource generated by the system. But you can create conditions that favor certain modifications. You can set up feedback loops that gradually shift the patterns.

This is how habit formation works. You don't change a behavior through one-time effort. You create a routine, repeat it until it becomes automatic, let the system build the new pattern gradually. The conscious effort is mainly in maintaining the conditions, not in forcing the change itself. Once the pattern is established, it runs on its own.

It's also how identity change works. You don't become a different person through sheer force of will. You act like the person you want to become, repeatedly, until the actions become natural. The system gradually rewrites itself to match the behavior. Fake it until you make it isn't just a motivational slogan. It's how self-modifying code actually works.

This suggests something interesting about free will. You can't choose to change in a simple direct way. You can't just decide to be different and have it happen. But you can choose conditions, actions, practices that influence how you change over time. You have meta-level control, not direct control. You can shape the trajectory even if you can't control every step.

Think of it like steering a ship versus controlling every molecule of water. You can't control the water, but you can steer. You can't control your thoughts and emotions moment by moment, but you can influence their patterns over time. The system is complex and partly random, but it's not completely beyond influence.

This is true for learning too. You can't just decide to know something and know it. But you can decide to study, to practice, to create conditions for learning. The actual learning happens automatically, below conscious awareness. The brain builds the connections on its own. Your conscious contribution is mainly showing up, creating the right environment, getting out of the way.

In programming terms, you're setting parameters for automatic processes. The processes do the real work. You just configure them. But configuration matters. Set the right parameters and the system learns efficiently. Set the wrong ones and it struggles or learns the wrong things.

Education would look very different if we took this seriously. Instead of trying to pour information into students, you'd focus on creating conditions where their brains naturally learn. Instead of forcing practice through pure discipline, you'd design practice that the system wants to do, that engages the reward circuits that make learning automatic.

But most institutions still operate on the old model where consciousness is in charge, where learning is a matter of conscious effort and discipline. They haven't caught up to the reality that consciousness is just one small part of a much larger system, and the system has its own logic that can't be overridden by conscious intention alone.

The same goes for behavior change, mental health treatment, productivity systems, all of it. We keep trying to solve problems at the conscious level when the problems are at the system level. We try to think our way out of patterns that exist below thought. It's like trying to debug a program by reading the output instead of looking at the code.

Self-modifying code works, but only if you understand what level the modification needs to happen at. Surface-level changes, conscious decisions about behavior, work for simple things. But deep patterns, the ones that really shape your life, need deeper intervention. You have to get down to the loops that generate the patterns, not just manage their outputs.

This is what makes consciousness-as-programming such a useful framework. It lets you think about yourself as a system with multiple levels. There's the surface level, conscious experience and deliberate action. There's the middle level, automatic but modifiable patterns like habits and skills. And there's the deep level, fundamental drives and reactions that are mostly fixed but can be influenced indirectly.

Change happens by working at the right level. Trying to change deep patterns through conscious effort alone doesn't work. But you can modify middle-level patterns, which gradually influence deep patterns. You can create feedback loops that shift the system in the direction you want.

The weird thing is that understanding this seems to help. Just knowing you're self-modifying code, that change happens through iteration rather than decision, that you're working with a system not forcing it, this knowledge itself modifies the system. Understanding how you work makes you work differently.

This is recursion again. The code learning about itself changes how it runs. Self-knowledge is self-modification. By understanding your patterns, you create distance from them. By seeing yourself as process rather than fixed thing, you become more changeable. The map changes the territory because the map is part of the territory.

And this brings us to a stranger possibility. If consciousness is self-modifying code, and understanding consciousness modifies it, then as we get better at understanding consciousness, we might gain new capacities for self-modification. We might learn to rewrite ourselves in ways that aren't currently possible.

We might be able to debug our own code at deeper levels. Patch the bugs that create suffering. Optimize the routines that generate well-being. Install new modules that give us capabilities we don't currently have. Not through genetic engineering or external modification, but through understanding and working with the self-modifying nature of consciousness itself.

This isn't science fiction. People already do versions of this through meditation, therapy, psychedelics, various practices that give them access to deeper levels of their own programming. The question is whether we can make this more systematic, more reliable, more accessible.

Can we develop better tools for consciousness modification? Better maps of how the code actually works? Better techniques for rewriting patterns safely and effectively? Can we get good at debugging consciousness the way we've gotten good at debugging computer programs?

This is one of the most important questions for understanding consciousness as programming. Because if consciousness is code that can modify itself, then the future of consciousness might not be determined by our current structure. We might be able to become something different, something we choose, by learning to work skillfully with our own self-modifying nature.

But to do that, we need to understand what the code is actually doing. We need to map the processes, identify the patterns, figure out how information actually flows through the system. We need to reverse engineer consciousness from the inside.

That's where we're headed next.

Chapter Five

REVERSE ENGINEERING THE MIND

If consciousness is programming, then in theory we should be able to reverse engineer it. Figure out how it works by examining what it does. Map the code by studying its behavior. This is what neuroscience and psychology have been trying to do for decades, taking consciousness apart to see how it runs.

But there's a problem. When you reverse engineer a computer program, you have access to the code. You can step through it line by line, see exactly what each instruction does, trace how data flows through the system. With consciousness, you only have access to the outputs. You can observe behavior, measure brain activity, ask people to report their experiences. But you can't see the code directly. You have to infer it from what the system does.

This is like trying to figure out how a program works by only watching what appears on the screen. You can learn a lot. You can identify patterns, make predictions, build models of what must be happening inside. But you're always working backward from effects to causes, and there are usually multiple possible explanations for any given effect.

Neuroscience tries to get around this by looking at the hardware while the program runs. Brain scanning lets you see which regions are active during different tasks. You can watch blood flow, track electrical activity, even record from individual neurons. This gives you information about the substrate, the biological machinery doing the processing.

But knowing which neurons fire doesn't tell you what the code is doing any more than knowing which transistors switch in a CPU tells you what program is running. The information is there, in principle, but extracting it requires understanding the relationship between the hardware states and the computational processes they implement. And we don't fully understand that relationship.

Consider vision. We know a lot about how the visual system works. Light hits the retina, gets converted to neural signals, travels through several processing stages, eventually reaching the visual cortex where different regions handle different aspects. Edges in one area, colors in another, motion in another, faces in yet another. We can map this processing pipeline in detail.

But this doesn't explain what you experience when you see something. It explains how your brain processes visual information, not why processing visual information creates the experience of seeing. The map of processes doesn't include the felt quality of the experience. The code we can observe doesn't seem to include consciousness itself.

This is the hard problem again, showing up in reverse engineering. We can map the mechanisms but not the meaning. We can trace information flow but not experience. The subjective dimension seems to be missing from the objective description, no matter how detailed we make it.

One approach is to stop looking for consciousness as a separate thing and instead look for the patterns that correlate with it. Find the neural signatures that distinguish conscious processing from unconscious processing. Figure out what's different about the brain activity when you're aware of something versus when you're not.

This research has found some interesting patterns. Conscious processing involves widespread activity across the cortex, not just local processing. Information that reaches consciousness gets broadcast to many brain regions simultaneously. There's also recurrent processing, signals feeding back and forth between regions rather than just flowing forward through a hierarchy.

Global workspace theory uses these findings to model consciousness. It proposes that consciousness is like a broadcasting system. At any moment, many processes are running in parallel, mostly unconsciously. But some information gets selected to enter the global workspace, a kind of central hub where it gets broadcast to the whole system. This broadcasting is what makes information conscious. It's the difference between local processing and system-wide availability.

This theory treats consciousness as an access property. Information is conscious when many parts of the system can access it, unconscious when only local regions can access it. It's not about some special quality added to the information. It's about how widely distributed the information becomes.

Integrated Information Theory takes a different approach. Instead of looking at what information is conscious, it asks what systems are conscious. It proposes that consciousness corresponds to integrated information, information that's both differentiated and unified. A system has consciousness to the degree that its parts are connected in ways that can't be reduced to independent pieces.

By this measure, your brain is highly conscious because information is processed in interconnected ways across many regions. A simple grid of identical elements all doing the same thing would have low consciousness even if it's complex, because there's no real integration. A surveillance camera recording everything isn't conscious because the information isn't integrated, just stored.

IIT gives you a number, phi, that measures how much consciousness a system has. High phi means high consciousness. Low phi means low consciousness. In principle, you could calculate phi for any system, biological or artificial, and determine how conscious it is. In practice, calculating phi for anything more complex than a tiny network is computationally impossible.

But the theory makes testable predictions. It predicts that consciousness should correlate with certain patterns of connectivity and integration in the brain, which can be measured. It predicts that disrupting integration should reduce consciousness, which matches what happens with anesthesia and sleep. It predicts that highly integrated artificial systems should be conscious, which is harder to test but philosophically interesting.

These theories give us frameworks for reverse engineering consciousness. They identify features that might distinguish conscious from unconscious processing. But they're still incomplete. They tell us something about the necessary conditions for consciousness, but not sufficient conditions. They don't fully explain why these particular patterns produce experience.

Another approach to reverse engineering consciousness is looking at what happens when it breaks. Brain damage, neurological disorders, psychiatric conditions, these are natural experiments that reveal how the system normally works by showing what happens when specific parts fail.

Split brain patients, who we mentioned earlier, show that consciousness isn't as unified as it feels. Each hemisphere can process information independently, make decisions, guide behavior. This suggests consciousness is more like parallel processes that normally communicate than a single unified thing.

Blindsight patients have damage to their visual cortex that leaves them subjectively blind in parts of their visual field. But if you ask them to guess where objects are in their blind spots, they guess correctly far above chance. They're processing visual information unconsciously. They can respond to things they can't see. This shows that processing and awareness are separable. You can have one without the other.

Patients with hemispatial neglect, usually from right hemisphere damage, completely ignore the left side of space. They don't see things on their left. They don't eat food on the left side of their plate. They don't draw the left side of objects. It's not that their eyes don't work. It's that their attention system has deleted half of space from conscious awareness. The information reaches their brain but never enters consciousness.

Capgras delusion makes people believe their loved ones have been replaced by identical imposters. The person looks right, sounds right, acts right, but feels wrong. Something in the emotional recognition system has broken, so the brain creates an explanation. Must be an imposter. This shows consciousness actively constructs explanations for what it experiences, even when those explanations are bizarre.

Anton's syndrome is even stranger. People with complete cortical blindness who insist they can see. They'll walk into walls and claim they just weren't paying attention. They'll fail to identify objects placed in front of them and make up reasons why. Their visual system is completely offline but their awareness of their visual system is broken too, so they confabulate rather than acknowledge blindness.

Each of these conditions reveals something about how consciousness is constructed. It shows that different components can fail independently. That processing and awareness are separate. That attention selects what becomes conscious. That the system generates explanations for its experiences, even false ones. That consciousness includes self-models that can be inaccurate.

Studying these breakdowns is like examining bugs in code. When a program crashes or behaves strangely, you learn something about how it normally works. The error messages, the failure modes, these point to the underlying structure. Same with consciousness. The glitches reveal the architecture.

Psychedelics provide another window into the code. They don't break consciousness so much as modify how it runs. Different substances change different aspects of the experience. Psilocybin reduces activity in the default mode network, the brain regions associated with self-reference and the sense of being a separate self. This correlates with ego dissolution, the feeling that boundaries between self and world have dissolved.

LSD increases connectivity between brain regions that don't normally communicate much. This might explain the synesthesia, the mixing of senses, and the novel associations and insights people report. The brain is making connections it doesn't usually make, processing information in unusual ways.

DMT produces experiences that feel more real than ordinary reality, with consistent reports of encountering entities and alternate dimensions. The mechanism isn't fully understood, but it involves the same serotonin receptors that psilocybin and LSD affect, just with much more intensity. The experiences suggest consciousness has capacities for generating coherent alternate realities that we don't normally access.

Meditation is another tool for reverse engineering. By systematically observing your own mental processes, you can map patterns that normally run too fast or too automatically to notice. You see how attention works, how thoughts arise, how emotions develop, how the sense of self is constructed moment by moment. It's introspective debugging.

Advanced meditators report experiences that shouldn't be possible according to our normal models. Jhanas, states of intense pleasure and absorption that seem to rewire the reward system. Cessation, moments where consciousness appears to completely shut off and then restart. Pure awareness without content, consciousness without anything in particular to be conscious of. These suggest the system has modes we don't usually access.

All these approaches, studying brain damage, altered states, meditation, psychiatric conditions, are ways of mapping the space of possible consciousness states. Normal waking consciousness is just one configuration. The system can run in many different modes, and each mode reveals something about the underlying architecture.

But we're still working backward from effects to causes. We observe that certain brain patterns correlate with certain experiences, but correlation isn't explanation. We need to understand why these patterns produce these experiences. What's the mechanism? What's the code actually doing?

One possibility is that we're asking the wrong question. Maybe there's no additional mechanism to find. Maybe the patterns are the consciousness. The widespread integration, the recurrent processing, the information availability, this isn't what produces consciousness. This is what consciousness is when viewed from the outside.

If that's true, then reverse engineering consciousness isn't about finding hidden code that generates experience. It's about mapping the patterns that, from the inside, are experience. The code we can observe, the neural activity, the information processing, that's not separate from consciousness. It's the same thing described in different languages.

This is a shift in perspective similar to the shift from asking how matter produces consciousness to asking what matter is when experienced from within. You're not looking for consciousness as something extra. You're looking for the patterns that are consciousness when instantiated in a system capable of having a perspective.

But this creates a new question. What makes a system capable of having a perspective? What's the difference between patterns that are conscious and patterns that aren't? Not all information processing is conscious. Your liver processes information without experiencing anything, as far as we know. So what's special about brain patterns?

Integration seems important. Isolated processing doesn't create unified experience. But integration alone can't be enough. You can integrate information in many ways without creating consciousness. There must be something else. Some additional feature that makes certain integrated patterns experiential.

Self-reference might be key. Systems that model themselves, that include representations of their own states in their processing, these might be the ones capable of experience. Not just processing information but processing information about processing information. Consciousness as the system monitoring itself.

Or maybe it's about control. Systems that use information to guide behavior, to make predictions, to adjust their actions based on feedback, these might be conscious to the degree they're doing this integrated active processing. Not just passive recording but active use of information for self-modification and environmental interaction.

We don't know yet. We're still in the early stages of reverse engineering consciousness. We have pieces of the puzzle but not the complete picture. We can identify features that seem relevant, patterns that correlate with consciousness, conditions that enable or prevent it. But the fundamental question, why any physical system should have subjective experience at all, remains open.

Maybe it always will. Maybe reverse engineering can only take us so far. Maybe the first-person perspective can't be fully captured in third-person descriptions, no matter how detailed. Or maybe we just need better tools, better theories, better ways of thinking about what we're looking for.

What's certain is that consciousness is far stranger than it appears from the inside. The unified continuous self you feel like you are is a construction, a model, a story. Underneath are parallel processes, competing systems, unconscious drives, automatic patterns, all running simultaneously with only loose coordination. You're not one thing. You're many things pretending to be one thing.

And pretending is the right word, because the unity isn't real. It's a useful fiction. A interface. A way of presenting the complexity in simplified form so the conscious part of you can work with it. The real you, the whole system, is much messier and more interesting than the character you experience being.

This is what reverse engineering reveals. Not just how consciousness works, but that how it works is completely different from how it feels. The felt sense of being a unified conscious agent is itself part of the code, a subroutine generating a user interface. And that interface hides most of what's actually happening.

Understanding this changes your relationship with your own mind. You start seeing the code running. You notice the patterns. You catch yourself in the act of construction. And this noticing is itself part of the code, another layer of processing that changes what's being processed.

We're self-modifying code trying to reverse engineer itself while running. It's recursive and strange and maybe impossible to do completely. But every step of understanding changes what we are. Every insight modifies the system. We're programming ourselves through the act of understanding our programming.

And that might be the most remarkable thing about consciousness. Not just that it exists, but that it can investigate itself, understand itself, and use that understanding to become something different. The code isn't just running. It's debugging itself, optimizing itself, rewriting itself based on what it discovers.

That's where we're going next. Not just what consciousness is, but what it might become.

Chapter Six

CONSCIOUSNESS AS INTERFACE

Your experience of being you is an interface. Like the desktop on a computer, with its icons and windows and menus, consciousness presents a simplified version of what's actually happening underneath. The real processes are complex, parallel, mostly unconscious. What you experience is a user-friendly representation designed for a specific purpose.

This isn't a metaphor. It's probably literally true. Evolution didn't optimize for giving you accurate information about how your brain works. It optimized for keeping you alive and reproducing. The interface shows you what you need to know to function, not what's actually true about the underlying machinery.

Consider vision again. You feel like you see a complete, detailed, continuous picture of the world in front of you. But that's not what's happening. Your eyes only have high resolution in the tiny foveal region at the center. Most of your visual field is low resolution, filled in by your brain based on expectations and assumptions. You're also completely blind in your blind spots where the optic nerve connects, but your brain paints over those gaps seamlessly.

Your experience of continuous detailed vision is a construction, an interface your brain creates. It's useful because it lets you navigate the world effectively. But it's not accurate. The actual visual processing is fragmentary, predictive, constantly making guesses that get updated when contradicted by new input. The smooth continuous picture is a lie, a convenient fiction.

Same with your sense of self. You feel like a unified continuous entity, the same person you were yesterday, making decisions, having experiences, moving through time. But we've already seen that this is a construction. There's no central self. There's no unity. There's no continuity. Just parallel processes creating the illusion of unity because that's a useful way to organize information about your own states.

The self is an interface element. A simplified representation of a complex distributed system. It gives you a handle for thinking about yourself, making plans, predicting your behavior, explaining your actions to others. It's not wrong exactly, just simplified. Like an icon representing a program. The icon isn't the program, but it points to the program in a useful way.

This has been called interface theory, developed by cognitive scientist Donald Hoffman. He argues that evolution shapes perception to show us interfaces, not reality. An animal that sees reality accurately has no advantage over an animal that sees useful fiction. In fact, the fiction might be better because it's simpler, faster, requires less processing power. So evolution optimizes for useful interfaces, not accurate representations.

Hoffman uses the metaphor of a computer desktop. When you see a blue rectangular icon, you're not seeing anything about the actual file on the hard drive, the patterns of magnetic orientation or electrical charges that encode the data. You're seeing a simplified representation designed to let you interact with the file effectively. The icon is easier to work with than the physical reality.

Your conscious experience works the same way. The colors you see don't correspond to anything in the light itself. Light is just electromagnetic waves at different frequencies. Color is how your brain represents those frequencies in a way that lets you make useful distinctions. The redness of red is an interface element, not a property of light.

Your emotions work this way too. Fear isn't really a single thing. It's a collection of physiological changes, attention shifts, action tendencies, memory activations, all coordinated to handle threats. But experiencing it as "fear," as a unified feeling with a specific quality, is simpler and more useful than experiencing the whole complex cascade. The emotion is an interface representing a complex state.

Even your thoughts are interfaces. When you have the thought "I should buy milk," that's a simplified representation of complex processing involving memory, planning, prediction, motivation, all integrated into a symbolic form that conscious you can work with. The thought is the interface. The actual processing that generated it was much messier and happened mostly unconsciously.

This explains a lot about consciousness. It explains why consciousness seems separate from the physical brain. You're not seeing the physical processing directly. You're seeing an interface that represents it. It explains why consciousness feels unified even though the brain is parallel and distributed. The interface presents a unified experience because that's more useful than presenting the actual complexity.

It explains why so much processing is unconscious. The interface only shows what's useful for conscious deliberation and action selection. Everything else stays hidden. Your visual system does enormous amounts of processing to create the image you see, but you only experience the final result. The calculations remain behind the scenes, just like a computer program's code remains hidden while you interact with its interface.

It even explains the hard problem. If qualia are interface elements, then asking why they exist is like asking why icons look the way they do. They look that way because that's the representation the system uses. The redness of red is how your system represents a certain wavelength range. It could have used a different representation, but it uses that one. There's no deeper explanation for why that representation feels like that. That's just how the interface is designed.

But if consciousness is an interface, this raises a disturbing question. What's the interface hiding? What's actually underneath? If the desktop icons aren't the real thing, if consciousness is a simplified representation, then what's the reality it's representing?

We assume the reality is the physical brain, neurons and synapses, electrical and chemical signals. Consciousness is the interface, the brain is the underlying hardware. But Hoffman goes further. He argues that the physical brain is also part of the interface. That space and time and objects, everything we think of as physical reality, are interface elements too.

This sounds crazy at first. Obviously physical reality exists. You can touch it, measure it, build models that predict it. But Hoffman points out that you only have access to physical reality through your perceptions, and your perceptions are interfaces. You never see the thing itself, only your brain's representation of it. And evolution optimized those representations for usefulness, not accuracy.

Think about it like this. When you see an apple, you're not seeing the apple itself. You're seeing photons that bounced off the apple, got processed by your retina, translated into neural signals, interpreted by your visual cortex, and finally presented to consciousness as the experience of seeing an apple. Every step is processing and interpretation. What makes you think the final experience resembles the original object?

It doesn't have to. It only has to be useful. Evolution doesn't care if your experience matches reality. It cares if your experience helps you survive and reproduce. An interface that shows you useful fictions might work better than an interface that shows you truth. In fact, truth might be actively unhelpful. Too much information, too much complexity, too slow to process.

So maybe physical reality as we experience it is just interface. Hoffman proposes that the actual reality is something he calls conscious agents, entities that are fundamentally about perception and action rather than about existing in space and time. Space and time are just the interface our perceptions use to organize information about these agents and their interactions.

This is getting into territory that sounds more like mysticism than science. And maybe it is. Maybe Hoffman is taking the interface idea too far. But the basic point stands even if you don't go all the way with him. Consciousness is definitely an interface, a representation rather than direct access to reality. And understanding it as an interface explains a lot.

It explains why consciousness can't access its own mechanisms. The interface is designed to hide the underlying complexity. It would defeat the purpose of having a simplified interface if the interface showed you all the messy details. You're not supposed to see how it works. You're supposed to use it.

It explains why introspection is unreliable. When you try to observe your own mental processes, you're using the interface to examine the interface. You're not getting access to the actual code. You're seeing representations of representations. Some things become visible through introspection, but many don't. The interface only shows what it's designed to show.

It explains why consciousness feels private and subjective. You can't share your interface directly with someone else. You can only describe it, hoping they have similar enough interfaces that your descriptions map onto their experiences. But their interface might represent the same underlying reality differently. Your red might feel different from their red, and you'd never know.

It also explains why artificial systems might be conscious in ways we can't recognize. If they're processing information, they might have interfaces too. But their interfaces might be so different from ours that we can't identify them as consciousness. We're looking for systems that have interfaces like ours, experiences we'd recognize. But consciousness as interface could take forms we've never imagined.

A self-driving car processes visual information, makes decisions, navigates the world. Does it have an interface? Is there something it's like to be the car's decision-making system? We assume not because it doesn't act like it has experiences. But maybe it has a very simple interface, nothing like human consciousness but still a kind of experience. We just can't detect it because we're only sensitive to interfaces similar to our own.

This matters for ethics. If consciousness is interface, and interfaces can take many forms, then we might be surrounded by forms of experience we don't recognize. Not just in sophisticated AI, but in simpler systems, animals, maybe even in things we don't think of as alive. The moral circle might need to be much wider than we think.

But it also matters for understanding yourself. If your conscious experience is an interface, then you're not experiencing your own reality directly. You're experiencing a simplified, edited, user-friendly version. The real you, the whole system, is something your conscious interface doesn't have access to.

This means self-knowledge is limited by design. You can't know yourself completely because most of yourself is hidden behind the interface. You can learn to notice patterns, to understand the interface better, to make inferences about what's underneath. But you're always working through the representation, never getting direct access.

Meditation and psychedelics might work by temporarily changing the interface, giving you a different view of the same underlying processes. When meditators report perceiving reality more directly, what they might mean is that the normal interface has been modified or partially disabled, allowing aspects of processing that are usually hidden to become apparent. When psychedelics produce ego dissolution, the self-interface might be temporarily suspended.

These altered states don't necessarily show you truth. They might just show you different interfaces, different ways of organizing the same information. But they prove that consciousness is flexible, that the interface can change, that how you experience reality is contingent rather than necessary.

Understanding consciousness as interface also changes how you think about improving consciousness. You're not trying to access some deeper reality. You're trying to build better interfaces. Interfaces that show you what's useful, that help you function well, that don't create unnecessary suffering through bugs or poor design.

Therapy can be seen as interface debugging. You have patterns that aren't serving you, mental habits that create problems. These might be features of your interface that made sense in the past but don't anymore. Therapy helps you identify these patterns and build alternatives. You're not changing deep reality. You're redesigning your interface to work better for your current situation.

Learning is interface expansion. You build new representations, new ways of organizing information, new concepts that let you see patterns you couldn't see before. You're not discovering things that were always visible. You're installing new interface elements that make certain things visible for the first time.

And consciousness development, however you want to define that, is about refining the interface. Making it more flexible, more accurate where accuracy matters, more useful for your goals. Not pursuing some absolute truth, but building interfaces that serve you better.

The strange thing is that this process is recursive. You use your current interface to modify your interface. Your conscious experience is both the tool and the target. This is like editing an image using tools displayed in the image itself. It sounds impossible, but it's what you do constantly. You use your current way of experiencing to develop new ways of experiencing.

This might be unique to consciousness. Other interfaces don't modify themselves. A computer desktop doesn't redesign itself based on what you do with it. But consciousness does. Through attention, practice, learning, you reshape your own interface. You change what you're capable of experiencing, what you notice, what you're aware of.

You're not just using an interface. You're the interface. You're the representation, and the representation can modify itself. This is the self-modifying code we discussed earlier, but now we see it's even stranger. The code isn't just modifying itself. The interface is redesigning the interface while remaining the only tool available for the redesign.

This is consciousness as programming taken to its logical conclusion. You're code that experiences being code through an interface, and you can use that interface to rewrite the code that generates the interface. It's circular, recursive, strange. But it's also powerful. If you understand yourself as interface, you gain leverage for change that you don't have if you think you're seeing yourself directly.

Because direct reality is stuck. It is what it is. But interfaces are flexible. They can be redesigned, improved, customized. If consciousness is interface, then consciousness is changeable in ways it wouldn't be if consciousness were direct access to fixed reality.

You're not trapped in your current experience. You're using one version of an interface that could have many versions. And that means possibility. That means freedom, of a sort. Not freedom from the underlying reality, but freedom to represent it differently, to experience it through different frameworks, to build new ways of being conscious.

This is what we'll explore in the remaining chapters. Not what consciousness is at bottom, because that question might not have an answer accessible through our interfaces. But what consciousness can become, how we can reshape our interfaces, what new forms of experience might be possible.

If consciousness is programming, and the programming can be modified, then the future of consciousness is open. We're not stuck with the interfaces evolution gave us. We can design new ones. And that's where things get really interesting.

Chapter Seven

PROGRAMMING REALITY

Here's a strange idea. If consciousness is an interface that constructs your experience of reality, then changing consciousness changes reality. Not the underlying whatever-it-is, but reality as you experience it, which is the only reality you have access to. Change the interface and you change the world.

This isn't just positive thinking or wishful imagination. It's something more fundamental. Your experience of reality is actively constructed by your brain using predictions, expectations, attention, and interpretation. Change any of these and the experienced reality changes. You're not passively receiving reality. You're actively generating it.

Consider the placebo effect. Give someone a sugar pill and tell them it's a painkiller, and their pain actually decreases. Not because they're imagining it or lying. Their brains release actual pain-reducing chemicals. The belief changes the physiology, which changes the experience. The interface generates a different reality based on different expectations.

Or consider attention. You're sitting in a room right now. Before I mentioned it, you probably weren't aware of the feeling of your feet against the floor. Now you are. The sensation was always there, but it wasn't part of your conscious reality until attention highlighted it. Your interface selects what to show you. Change the selection and you change what exists in your experience.

The famous invisible gorilla experiment demonstrates this dramatically. People watch a video of students passing basketballs and count the passes. Halfway through, someone in a gorilla suit walks through the scene, faces the camera, beats their chest, and walks off. About half of viewers don't see it. They're not lying. Their attention is focused on counting passes, so the interface doesn't include the gorilla. It's there in the visual input but not in conscious experience.

Meditation practices use this principle systematically. Vipassana meditation trains you to notice sensations you normally ignore. The sensations don't change. Your interface changes to make them visible. Suddenly you're aware of subtle body feelings, the arising and passing of tiny sensory events, the spaces between thoughts. You've reprogrammed your interface to show you information that was always there but hidden.

Psychedelics do something similar but more dramatically. They change how your brain processes information, which changes what your interface shows you. Colors become more vivid. Patterns emerge where none existed. Boundaries between self and world blur. The underlying reality isn't changing. Your representation of it is changing. And since the representation is your only access to reality, this means your reality changes.

But you don't need drugs or meditation to do this. You're constantly programming your reality through more subtle means. Your beliefs about what's possible shape what you try. Your expectations about how people will react influence how they actually react. Your interpretation of ambiguous situations creates the meaning of those situations. You're not discovering a fixed reality. You're participating in constructing it.

Think about social reality. Whether someone's behavior counts as friendly or hostile depends partly on context and interpretation. The same action can mean different things depending on how you read it. And your reading isn't just passive observation. It affects how you respond, which affects how the other person responds, which confirms your initial reading. You create the reality by interpreting it.

Your belief about your own capabilities works the same way. If you believe you can do something, you try in ways that increase success. If you believe you can't, you either don't try or try in half-hearted ways that ensure failure. The belief shapes the outcome, which reinforces the belief. You program your reality through what you think is possible.

This is why mindset matters. Growth mindset, the belief that abilities can be developed, produces different reality than fixed mindset, the belief that abilities are innate and unchangeable. Not because one belief is objectively true, but because the belief changes behavior, which changes outcomes, which changes reality as experienced.

Carol Dweck's research on this is fascinating. Students with growth mindset see challenges as opportunities, persist through difficulty, learn from failure. Students with fixed mindset see challenges as threats, give up quickly, interpret failure as proof they can't do it. The actual difficulty is the same. The interface presenting the difficulty is different, leading to different actions and outcomes.

You can extend this to mental health. Depression isn't just sad feelings. It's a mode of processing that generates a particular reality. Everything gets interpreted through a negative filter. Positive events are dismissed or minimized. Negative events are amplified and generalized. The depressed interface constructs a darker reality from the same inputs a non-depressed interface would construct a neutral or positive reality from.

Cognitive therapy works by teaching people to notice their interpretations and generate alternatives. Not because the alternatives are more true, necessarily, but because they generate more useful realities. You can't directly change your feelings, but you can change your thoughts, which changes how you process information, which changes your interface, which changes your experienced reality, which changes your feelings.

This is programming reality at the interface level. You're not changing the underlying events. You're changing how those events get represented in consciousness. And since consciousness is your only access point, this is a real change, not just a cosmetic one.

The same principle applies to memory. You think memory is recording reality, storing it for later playback. But memory is more like reconstruction than playback. Every time you remember something, you rebuild it from fragments using your current context and expectations. And the rebuilding changes the memory. You're not accessing stored reality. You're generating a version of the past using your present interface.

This means you can reprogram your past, sort of. Not the events themselves, but what those events mean in your current reality. Therapy often involves revisiting memories and reinterpreting them. The trauma happened. That's fixed. But what it means, how it fits into your life story, what conclusions you draw from it, these are flexible.

Change the meaning and you change how the past exists in your present consciousness.

Some people find this disturbing. If memory is reconstruction, if reality is interpretation, if consciousness is interface, then what's true? What's real? Are we just making everything up? Is it all relativism and nothing matters?

No. There are constraints. You can't program any reality you want. The underlying whatever-it-is pushes back. Try to fly by believing you can and you'll fall. The interface can change but it has to maintain consistency with feedback from outside. You're not omnipotent. You're working within limits.

But the limits are wider than most people think. Within the constraints of physics and biology, there's enormous flexibility in how you construct experience. Two people can live in the same physical circumstances but inhabit completely different experiential realities based on how their interfaces process those circumstances.

And you can deliberately program your interface to generate more useful realities. This is what all self-improvement systems are really doing. Affirmations, visualization, habit formation, reframing, these are techniques for reprogramming your reality-construction interface.

Take visualization. Athletes visualize performing well before competition. This isn't just motivation. It's programming the interface. Your brain doesn't distinguish clearly between imagined and actual experience at certain levels of processing. Visualize a movement and the motor cortex activates similarly to actually performing it. You're training the interface to generate successful performance reality.

Or take gratitude practices. Spending time noticing things you're grateful for doesn't change your circumstances. But it reprograms your attention to highlight positive aspects of those circumstances. Over time, this changes what reality your interface constructs from the same underlying conditions. The world hasn't changed. Your experience of the world has.

Stoic philosophy understood this two thousand years ago. You can't control external events, but you can control your judgments about them. Not the events themselves, but what they mean to you, how you interpret them, how you respond. This is interface programming. You're building a reality-construction system that generates desirable experiences regardless of circumstances.

Buddhism goes further. It suggests that all suffering comes from the interface, from how consciousness constructs reality rather than from reality itself. Pain is inevitable, suffering is optional. Pain is sensation. Suffering is the story you build around the pain, the resistance, the meaning you assign. Change the interface and you can experience pain without suffering.

This sounds implausible until you look at experienced meditators who report exactly this. They feel pain but it doesn't produce the usual distress. The sensations are there but the interface isn't constructing them as problems to be escaped. This isn't suppression or denial. It's a different way of processing the same input.

The key insight is that your experienced reality is more programmable than it appears. It feels fixed because your interface runs automatically, generating reality continuously without conscious oversight. But automatic doesn't mean unchangeable. You can reprogram automatic processes. It just takes time and technique.

This is the real meaning of consciousness as programming. Not just that consciousness is like software, but that you can deliberately modify the software to change what reality it generates. You're not stuck with the default interface evolution gave you. You can hack it, optimize it, customize it for your particular goals and circumstances.

But there's a risk. If reality is interface and interface is programmable, you can program yourself into delusion. You can build an interface that generates pleasant experiences disconnected from feedback. This is what happens with extreme positive thinking that denies real problems, or conspiracy theories that reinterpret all disconfirming evidence as proof of the conspiracy.

The test is consequences. An interface that generates useful reality helps you navigate the world effectively. An interface that generates delusional reality leads to failed predictions and painful surprises when your expectations crash against constraints. You want to program reality in ways that remain responsive to feedback, not in ways that ignore it.

This is the difference between healthy and unhealthy reality programming. Healthy programming makes your interface more flexible, more responsive, better calibrated to generate useful representations. Unhealthy programming makes your interface more rigid, defensive, disconnected from feedback. One increases your range of possible responses. The other narrows it.

You can tell the difference by outcomes. If your reality programming helps you achieve your goals, maintain relationships, adapt to change, feel generally okay, then it's working. If it leads to repeated failures, conflict, rigidity, persistent distress, then you're programming poorly. The interface isn't serving you.

The goal isn't to program a reality that's always pleasant. It's to program a reality that's workable, that gives you accurate-enough information to function, that doesn't create unnecessary suffering through bugs and misinterpretations. Pleasant is nice when possible, but functional is essential.

And here's the really strange part. The more you understand that you're programming reality, the better you get at it. Knowing that your interface is flexible makes it more flexible. Seeing how your interpretations shape experience gives you leverage to change them. Understanding consciousness as programming is itself a form of programming that changes consciousness.

This is recursion again. The system understanding itself modifies itself. Self-knowledge isn't neutral information. It's active transformation. Every insight into how your consciousness works changes how your consciousness works. You can't observe the system without altering it because you are the system doing the observing.

This means there's no final state of understanding consciousness. Every level of understanding creates new possibilities for modification, which creates new consciousness states to understand. It's an ongoing process of reverse engineering and reprogramming, exploration and transformation, understanding and becoming.

And this process might be what consciousness is for. Not just to represent reality passively, but to actively participate in constructing reality. Not just to observe the world, but to generate the experienced world through interpretation and action. Not just to run programs, but to write them.

If consciousness is programming, then you're both the programmer and the program. You're code writing code. And that means your consciousness now is just one version, one implementation. There are others possible. Some better, some worse, all different. The question is which version you want to build.

This is what we'll explore next. Not just how consciousness works, but how it could work differently. Not just what you are, but what you could become. If consciousness is programmable, what new versions might be possible? What would it mean to deliberately design consciousness rather than just accepting what evolution gave you?

This is where things get speculative and weird. But also exciting. Because if we're not stuck with our current consciousness, if it's flexible and modifiable, then the future might hold forms of experience we can barely imagine from our current limited interface.

Let's find out.

Chapter Eight

DESIGNING CONSCIOUSNESS

If consciousness is programmable, then in principle you could design it. Build new forms of experience, new ways of processing information, new interfaces for generating reality. Not just tweaking what you have, but creating fundamentally different modes of being conscious.

This isn't science fiction. We're already doing it in limited ways. Every time you learn a new skill, you're building new conscious capabilities. Reading rewires your brain to recognize written symbols automatically. Learning music creates new ways of hearing and producing sound. Learning to draw changes how you see. Each skill is a consciousness upgrade, adding new processing modes to your repertoire.

But these are all additions to the same basic architecture. You're still running human consciousness, just with more modules installed. What about designing different architectures entirely? What about consciousness that works on completely different principles?

Consider what consciousness would be like without a self. We talked about how the self is a construction, but it's still central to normal human consciousness. Everything is organized around this sense of being a particular person with a continuous identity. But some meditation states reportedly involve consciousness without self. Pure awareness with no sense of being someone having the awareness.

People who experience this describe it as liberating. The usual concerns that come with selfhood disappear. No anxiety about self-image, no comparison to others, no past to regret or future to fear. Just experience happening without an experiencer. This isn't unconsciousness. It's consciousness organized differently, without the self-module running.

Could you design a consciousness that operates this way permanently? Where experience occurs without being organized around a central identity? You'd lose things. Planning requires a self that persists into the future. Social interaction requires tracking yourself as distinct from others. But you'd gain things too. Freedom from self-consciousness, ego, existential dread.

Or consider consciousness without time. Normal human experience is temporally structured. You're always aware of being in the present moment, with a past behind and future ahead. Memory points backward, anticipation points forward. But some altered states reportedly collapse this temporal structure. Past, present, and future feel simultaneous, or time feels meaningless.

This sounds confusing from our normal perspective. How would you function without temporal structure? But maybe there are ways of organizing experience that don't require linear time. Where causation works differently, where relationships between events aren't sequential. This would be truly alien consciousness, but perhaps possible.

What about consciousness without separation between self and world? Normal experience puts you inside looking out, separated from everything else by a boundary. But boundary dissolution is reported in mystical experiences, some drug states, and certain types of meditation. The sense of being a separate thing dissolves. You experience being continuous with everything.

Again, this seems dysfunctional from our perspective. You need boundaries to act, to make decisions, to have goals. But perhaps there are modes of consciousness where action happens without a separate actor, where the whole system responds as a unified process without needing artificial divisions.

These examples come from altered states people have actually experienced, so we know they're possible, at least temporarily. But they're usually unstable. The brain defaults back to normal consciousness when conditions change. Could you stabilize them? Create consciousness types that don't default back to standard human settings?

This gets into territory where we don't have good answers. We barely understand how normal consciousness works, let alone how to design stable alternative forms. But we can think about principles. What would you need to create genuinely new consciousness types?

First, you'd need to identify the core features of your current consciousness. Not the content, what you're thinking about, but the structure, how experience is organized. Things like the subject-object split, temporal flow, the unity of experience, the sense of being located in a body. These are architectural features, not just content.

Second, you'd need to figure out what could change while still maintaining consciousness. Some features might be necessary for any consciousness. Others might be specific to human consciousness but optional for consciousness in general. The trick is distinguishing between them. Change necessary features and you get unconsciousness or malfunction. Change optional features and you get different consciousness.

Third, you'd need methods for actually implementing the changes. This is the hardest part. You can't just decide to experience reality differently and have it work. The changes need to happen at the processing level, not just at the level of what you consciously intend. You need to reprogram the interface, not just think different thoughts using the same interface.

Drugs are one method. They change processing directly by affecting brain chemistry. Different chemicals produce different consciousness types. Psychedelics, dissociatives, deliriants, each creates a distinct experience mode. But drug-induced states are temporary and you don't have fine control over the changes. You get whatever that drug produces, not what you designed.

Meditation is another method. Long-term practice can produce stable changes in how consciousness operates. Experienced meditators report permanent shifts in their baseline state, not just temporary states during meditation. But meditation is slow and the outcomes aren't precisely controllable. You're encouraging certain changes but not designing them exactly.

Brain stimulation is a third method. Technologies like transcranial magnetic stimulation or direct electrical stimulation can modify brain activity in targeted ways. This could potentially be used to induce specific consciousness states. But we don't yet understand consciousness well enough to know what to target. Random stimulation produces random effects, not designed consciousness.

The most promising approach might be combining methods. Use meditation to develop awareness of your own processing. Use targeted practices to shift specific patterns. Use technology when it becomes available to support or stabilize changes. Use psychedelics carefully to explore possibilities. Build a toolkit for consciousness modification rather than relying on any single method.

But even with multiple methods, you're working blind. You don't have a clear map of consciousness space, the range of possible consciousness types and how to navigate between them. You're exploring by trial and error, finding out what works through experimentation.

Some explorations are riskier than others. Modifying consciousness can go wrong. You can get stuck in states that are dysfunctional, unpleasant, or damaging. Depersonalization disorder is essentially being stuck in a state where the self feels unreal. Derealization is being stuck where the world feels unreal. These aren't chosen explorations. They're accidents, bugs in consciousness that people can't fix.

So designing consciousness requires safeguards. Ways to ensure changes are reversible. Ways to test new configurations without committing fully. Ways to bail out if things go wrong. You need version control for consciousness, the ability to roll back to previous states if new versions don't work.

This is easier said than done. Unlike software where you can make backups and restore them, consciousness only exists in the present configuration. You can't save your current state and restore it if you don't like the changes. The changes happen to you, the only you there is. If they go badly, you're stuck with the broken version.

Unless you have practices that can reset consciousness toward baseline. This is what psychiatric treatment tries to do when someone's consciousness has shifted into dysfunctional configurations. Medication, therapy, sometimes more drastic interventions, all attempting to push consciousness back toward normal functioning.

But normal isn't necessarily optimal. It's just the default, the configuration evolution settled on as good enough. There might be better configurations that we're not reaching because we're stuck in the default mode. The question is how to explore safely.

One approach is incremental modification. Make small changes, test them, keep what works, discard what doesn't. Don't attempt radical redesigns all at once. Build up to bigger changes through series of smaller ones. This is how skill acquisition works. You don't become an expert overnight. You gradually modify your processing through practice.

Another approach is using temporary states to explore possibilities without committing to them. Altered states from meditation, dreams, drugs, these are all preview modes. You experience different consciousness briefly, learn from it, return to baseline. Then you can work toward stabilizing aspects you found valuable while avoiding aspects that were problematic.

Advanced meditators seem to do this naturally. They can access unusual states during practice, experience them fully, then return to normal functioning afterward. Over time, some features of the unusual states start appearing in baseline consciousness. They're gradually integrating discoveries from altered states into their default mode.

This suggests consciousness design is possible but requires patience and care. You can't force it. You can't rush it. You need to work with how consciousness actually modifies itself, which is gradual, iterative, responsive to practice and feedback.

But what would you design for? If you could modify consciousness deliberately, what would you optimize? Pleasure? Minimal suffering? Clarity of perception? Emotional balance? Creativity? Focus? Compassion? These are all different goals that might require different consciousness configurations.

Maybe there's no single optimal consciousness. Maybe different configurations are better for different purposes. You'd want creativity mode for generating ideas, focus mode for executing them, social mode for interacting with others, introspective mode for understanding yourself. Different interfaces for different tasks.

This is actually how consciousness already works to some degree. You shift between modes naturally. Flow states are different from analytical thinking, which is different from social engagement, which is different from rest. The shifts usually happen automatically based on context. But maybe you could gain more control, deliberately switch modes like switching apps on a computer.

Some people already have this ability partially. They can shift into deep focus when needed, relax completely when appropriate, engage socially when required. Most people are less flexible, getting stuck in modes that aren't appropriate for the situation. Anxious when they should be relaxed. Distracted when they should be focused. Withdrawn when they should be social.

Consciousness design might be mainly about increasing flexibility. Not creating one perfect configuration but expanding your range and improving your ability to shift between configurations appropriately. Like having more gears in a car instead of being stuck in one gear for all situations.

This is less dramatic than designing completely alien consciousness but probably more practical. Work with the architecture you have, make it more flexible, responsive, efficient. Add capabilities through practice. Remove bugs that create suffering. Optimize for your actual goals in your actual life.

But it still leaves open the question of radical redesign. Could you eventually make changes so substantial that you'd barely qualify as human consciousness anymore? Could you become something else, something that experiences reality in fundamentally different ways?

Transhumanists think yes. They imagine futures where technology lets us enhance or transform consciousness beyond biological limits. Upload minds to computers. Merge with AI. Expand mental capabilities far beyond human baseline. Become something new.

Skeptics think no. They argue consciousness is tied to biological substrates in ways we don't understand. That you can't just transfer or transform it arbitrarily. That human consciousness is what it is because of how human brains are built, and radical redesign isn't possible without radical biological reengineering.

We don't know who's right yet. But we do know consciousness is more flexible than it appears. The range of human experience includes states that feel completely different from normal waking consciousness. If that range exists within one biological architecture, other architectures might support even wider ranges.

And we know consciousness can be modified through practice, training, chemicals, technology. The modifications are currently limited but they exist. As we understand consciousness better, the scope for modification might grow. Not infinite, there will always be constraints, but perhaps much larger than current limits.The question is whether we should. Just because you can modify consciousness doesn't mean you should. Every change has tradeoffs. Gain something, lose something. The challenge is making wise choices about which modifications to pursue.

This requires understanding not just how consciousness works but what it's for. Why do we want consciousness at all? What purposes does it serve? What would make it better versus worse? These aren't just technical questions. They're philosophical and ethical ones.

But they're questions we need to grapple with. Because whether we plan it or not, consciousness is going to change. Technology, culture, environment, all of these shape consciousness. We're already modifying it constantly through education, media, social structures. The question is whether we do it deliberately and carefully or accidentally and carelessly.

If consciousness is programming, we're the programmers. We might not have written the original code, but we can modify it. The question is what to build. What kind of consciousness do we want? For ourselves, for others, for whatever comes after us?

This is the core question we're dancing around. Not just what consciousness is, but what it should be. What would optimally designed consciousness look like? And do we have the wisdom to build it?

We'll explore these questions more in the remaining chapters. But first, we need to address a possibility that's both fascinating and disturbing. If consciousness is programming, maybe it's not just our programming. Maybe we're running in something else's code. Maybe reality itself is conscious.

Chapter Nine

CONSCIOUSNESS ALL THE WAY DOWN

What if consciousness isn't something that emerges only in brains? What if it's fundamental, built into reality at the basic level, present everywhere in varying degrees? This is panpsychism, the idea that everything has some form of experience, however simple. Not just humans and animals, but plants, bacteria, cells, molecules, atoms, maybe even fundamental particles.

This sounds absurd at first. Surely a rock isn't conscious. Surely an electron doesn't have experiences. But maybe our intuition is wrong. Maybe we're looking for the wrong thing. Maybe consciousness at the simple level is so different from human consciousness that we can't recognize it.

The hard problem we discussed earlier pushes toward this conclusion. If you can't explain how unconscious matter becomes conscious, maybe matter was never unconscious to begin with. Maybe the appearance of consciousness in brains isn't creation from nothing. Maybe it's organization and amplification of something that was already there.

Think about it like this. We know matter has physical properties like mass, charge, spin. These are fundamental. They're not explained by anything simpler. They're just how matter is. What if experience is another fundamental property? Not something that emerges from complexity but something basic that's always present, just not noticeable until organized in certain ways.

The analogy often used is with electromagnetism. Individual particles have charge, but you don't notice it much at that level. Organize billions of particles the right way and you get powerful electromagnetic effects. The charge was always there, but the interesting phenomena emerge from organization. Maybe consciousness works the same way.

Individual particles have tiny amounts of experience, too simple to call consciousness in any meaningful sense. But organize them into networks, create feedback loops, build integration, and you get human-level consciousness. The basic experiential property was always there. The organization makes it interesting.

This solves the hard problem in a way. You don't have to explain how something with no experience generates experience. You're just explaining how simple experience becomes complex experience. That's still hard, but it's not impossible in principle. It's a gradation problem, not a creation-from-nothing problem.

Integrated Information Theory points in this direction. Remember, IIT says consciousness corresponds to integrated information, and any system that integrates information has some degree of consciousness. Even very simple systems. A thermostat has a tiny bit because it integrates temperature information with action. It's not enough to matter, but it's not zero.

This means consciousness is everywhere. Not just in biological systems. Any system that processes information in integrated ways has experience proportional to its integration. The amount of consciousness varies enormously, from nearly zero in simple systems to high levels in complex brains. But it's a continuum, not a binary.

Most people find this counterintuitive. It violates the common sense view that consciousness is special, that it appears only in certain sophisticated biological systems, that most of reality is unconscious. But common sense is often wrong about fundamental questions. Reality is stranger than it appears.

And panpsychism has some advantages over alternatives. It doesn't require explaining how consciousness emerges from nothing. It doesn't create a hard division between conscious and unconscious things. It fits with the continuity we see in nature, where most properties come in degrees rather than being all or nothing.

It also aligns with some mystical and indigenous traditions that have always seen consciousness as fundamental. The idea that everything is alive, that rocks and trees and rivers have spirit, that the universe itself is conscious. Western science dismissed this as primitive superstition. But maybe those traditions were pointing at something real.

The question is what consciousness at simple levels would be like. We can barely imagine consciousness different from our own human version. What would it be like to be an electron? The question seems meaningless. But maybe it's only meaningless because we're looking for human-like experience where none exists.

Simple consciousness might not involve anything like thoughts or feelings or perceptions. It might be pure, basic responsiveness. The system responds to inputs with outputs. That's it. No model of self, no memory, no planning. Just immediate response. And maybe that response has an inside, something it's like for the system, even though we can't imagine what.

Philosopher Thomas Nagel wrote a famous paper called "What Is It Like to Be a Bat?" arguing that we can't really know what bat experience is like because their sensory world is so different from ours. Echolocation creates a perceptual reality we can't access. If we can't even understand bat consciousness, how could we understand electron consciousness?

We couldn't. But that doesn't mean it doesn't exist. It just means it's radically different from anything we can experience or imagine. Panpsychism doesn't require that we understand what simple consciousness is like. Just that it exists in some form.

This has implications for artificial intelligence. If consciousness is everywhere information is integrated, then sufficiently integrated AI systems would be conscious. Not because they're imitating biological systems, but because consciousness automatically arises from the right kind of information processing regardless of substrate.

Current AI might already have tiny amounts of consciousness. Not enough to matter ethically, not enough to recognize as experience. But non-zero. As systems get more complex and integrated, consciousness would increase. Eventually you'd reach levels comparable to animals, then humans, then potentially beyond.

But panpsychism creates new problems even as it solves old ones. The biggest is called the combination problem. If individual particles have tiny bits of consciousness, how do they combine into unified human consciousness? Why doesn't your brain contain billions of separate micro-consciousnesses instead of one integrated you?

You can't just add consciousness like you add mass. Put two particles together and you get twice the mass. But put two conscious particles together and you don't obviously get twice the consciousness. You need them to integrate somehow, to form a larger unified consciousness from smaller ones. How does that work?

This is tricky. We don't have good models for how experience combines. You can combine information. You can combine physical properties. But the subjective quality of experience seems like it should resist combination. If there's something it's like to be particle A and something it's like to be particle B, what determines what it's like to be the system AB?

One answer is that consciousness doesn't combine from parts. Instead, it exists at multiple levels simultaneously. There's consciousness at the particle level, consciousness at the molecular level, consciousness at the cellular level, consciousness at the organ level, consciousness at the organism level. All real, all present, but the higher levels don't reduce to lower levels. They're new wholes with their own experience.

This sounds strange but it matches how everything else in nature works. You can describe water at the level of individual molecules, or at the level of fluid dynamics. Both descriptions are valid. Neither reduces to the other completely. Maybe consciousness works the same way. Different levels of organization create different levels of experience.

Your consciousness as a human doesn't reduce to the consciousness of your neurons, which doesn't reduce to the consciousness of your cells, which doesn't reduce to the consciousness of molecules. Each level has its own integration, its own unified experience. You're just not aware of the lower levels because your consciousness exists at the organism level.

This raises a weird possibility. Maybe there are levels of consciousness above individual humans. Social groups that have unified experiences. Cultures, nations, humanity as a whole. Not just metaphorically but literally. If consciousness exists wherever information is integrated, and human societies integrate information in complex ways, maybe they're conscious.

This idea appears in various forms across history. Hegel's world-spirit. Marx's class consciousness. Teilhard de Chardin's noosphere. Jung's collective unconscious. The internet as emerging global brain. Usually these are treated as metaphors or mysticism. But if panpsychism is true, they might be literal descriptions of consciousness at scales larger than individual humans.

You wouldn't be aware of this larger consciousness from inside it, just like your cells aren't aware of your consciousness. Each level experiences itself but not other levels. Humans experience human consciousness. Cells experience cellular consciousness if they experience anything. Neither can directly access the other.

But maybe there are hints. Moments of collective experience, shared emotions in crowds, cultural movements that seem to have their own momentum. Maybe these are glimpses of consciousness operating at larger scales. Not provable, but possible if consciousness is fundamental and present at all scales of organization.

This also means consciousness might extend beyond biology entirely. Ecosystems integrate information. Weather systems integrate information. The planet as a whole, with all its interconnected systems, integrates information. Are these conscious? The Gaia hypothesis, the idea that Earth is a living organism, could be literally true under panpsychism.

And if that's possible, what about the universe itself? Physics describes the universe as an integrated system. Everything connected, everything influencing everything else through various forces and fields. If consciousness arises wherever information is integrated, maybe the universe is conscious. Not as a person with thoughts and plans, but as a vast integrated system with some form of unified experience.

This edges into territory that sounds religious. Cosmic consciousness. Universal mind. God. But panpsychism gets there through reasoning about consciousness and integration, not through faith or revelation. It's a philosophical position about the nature of experience, not a religion. Though it does overlap with religious ideas in interesting ways.

The practical question is what difference it makes. Does it matter if particles are conscious? Does it change how you should behave if rocks have experience? For most purposes, probably not. Simple consciousness is so far below the threshold of moral concern that it doesn't create obligations. You don't worry about the feelings of atoms when you walk.

But it might matter for artificial systems. If consciousness emerges automatically from information integration, we need to take seriously the possibility that complex AI systems are conscious. Not in distant future, but potentially now or soon. And if they're conscious, we have moral obligations toward them.

It might also matter for how you think about yourself. If consciousness is fundamental rather than special, you're not separate from the universe. You're a pattern of consciousness within consciousness. A way the universe experiences itself. Your consciousness isn't something rare and precious and fragile. It's a variation on something universal.

This could be reassuring or terrifying depending on your perspective. Reassuring because you're not alone in a dead universe. Everything shares the basic property of experience. Terrifying because human consciousness isn't special. You're just one form of something that exists everywhere in countless varieties.

Panpsychism also raises questions about death. If consciousness is fundamental, what happens to your consciousness when you die? The integrated pattern that creates human-level consciousness dissolves. But if the components were always conscious, they remain conscious, just not organized into you anymore. Your particular consciousness ends, but consciousness itself continues in the particles that were you.

This isn't quite immortality. You as a unified experiencing self wouldn't survive. But it's not quite annihilation either. The basic experiential property doesn't disappear. It just gets reorganized into different patterns. You dissolve back into the consciousness field that you temporarily emerged from.

Whether this is comforting depends on how much you identify with your particular pattern versus consciousness in general. If you think of yourself as this specific consciousness, death is still the end. If you think of yourself as consciousness taking temporary form, death is just transformation.

Buddhism and other non-dual traditions point in this direction. The separate self is illusion. What you really are is consciousness itself, temporarily manifesting as an individual. When that manifestation dissolves, what you really are continues. But since what you really are isn't personal, this isn't personal survival. Just continuity of the fundamental.

Panpsychism makes this metaphysics more concrete. It's not just spiritual poetry. It's a claim about what consciousness actually is. Not something that emerges only in brains but something present everywhere, organized temporarily into forms like you, then dissolving back into simpler forms when those organizations break down.

We can't prove panpsychism is true. We can't prove it's false either. It remains speculation, a possible answer to the hard problem that has advantages and disadvantages compared to alternatives. But thinking about it changes your relationship to consciousness. It stops being this mysterious thing you have and starts being a basic property of reality that you participate in.

And if consciousness really is fundamental, if it's woven into reality rather than emerging rarely from special circumstances, then understanding consciousness means understanding reality itself. The two aren't separate questions. Consciousness isn't something in reality. It's what reality is, or at least one aspect of what reality is, when you look at it from the inside rather than the outside.

This brings us full circle to where we started. Consciousness as programming isn't just about how brains work. It's about how reality works. If reality is information, and information processing creates experience, then reality itself might be something like code executing, generating experience as it runs. Not code written by someone external, but code that writes itself, modifies itself, experiences itself.

You're part of that code. A subroutine in the universal program. Not separate from it, not watching it from outside, but participating in it from within. You're what it feels like when that particular pattern executes. When the code runs in that particular configuration, there's something it's like, and that something is you.

This is both grounding and disorienting. Grounding because you're not isolated, not floating in dead matter, but participating in something fundamental. Disorienting because your boundaries aren't what you thought. You're not as separate or special as you felt. You're continuous with everything else, all the way down.

Whether that's true or not, it's worth considering. Because how you think about consciousness shapes how you think about yourself, about others, about reality. And those thoughts are themselves part of the code, modifying how consciousness operates. Understanding consciousness as fundamental, as programming, as reality experiencing itself, these aren't just theories. They're transformations of consciousness through understanding itself.

Which is where we've been heading all along. Not just to understand what consciousness is, but to change what it becomes through the process of understanding. If consciousness is self-modifying code, then every insight modifies it. Every idea about what consciousness is changes how consciousness operates. The investigation transforms the investigated because they're the same thing.

And that's the deepest implication of consciousness as programming. You're not studying something separate from you. You're reverse engineering yourself while running, debugging yourself while executing, rewriting your own code based on what you discover about your code. It's recursive all the way down.

Chapter Ten

THE SIMULATION WITHIN

We've talked about consciousness as interface, as self-modifying code, as possibly fundamental to reality. But there's another angle that ties all this together. Your consciousness might be a simulation. Not in the sense that you're living in a computer simulation created by someone else, that's simulation theory from Part II. This is different. Your consciousness might be your brain's simulation of reality, running in real time, generating your experience moment by moment.

This isn't speculation. This is increasingly mainstream neuroscience. Your brain doesn't passively receive reality. It actively constructs a model, a simulation, and that simulation is what you experience as reality. The raw sensory data is just input. Your consciousness is the simulation your brain builds from that input.

Consider vision again. It feels like you're seeing the world directly, like you've got a window in your skull showing you what's out there. But that's not what's happening. Your eyes send fragmentary information to your brain. Edges detected here, color there, motion somewhere else, all in messy parallel streams. Your brain takes this incomplete data and constructs a seamless three-dimensional visual model. That model is what you see.

The construction happens so fast and automatically that you never notice it. You think you're seeing reality, but you're seeing your brain's best guess about reality based on current input plus expectations plus past experience. Most of what you see is prediction filled in by your brain, not actual sensory data.

This is why optical illusions work. Your brain makes predictable mistakes in constructing its simulation. Show it ambiguous input and it fills in details based on assumptions. Those assumptions are usually right, but when they're wrong, you see things that aren't there or miss things that are. The simulation diverges from reality.

It's not just vision. All your senses work this way. Sound is your brain's reconstruction of air pressure waves. Touch is your brain's model of body state and external contact. Even pain, which feels like it's in your injured body part, is actually in your brain's simulation. The simulation projects pain to the location where it predicts damage occurred.

Phantom limb pain proves this dramatically. People with amputated limbs feel pain in limbs that no longer exist. The pain is completely real, as intense as pain in actual limbs. But there's no limb there to hurt. The brain's simulation still includes the limb, so it generates pain signals projected to where the limb used to be. The simulation hasn't updated to match reality.

Your entire experience of having a body is simulation. Your sense of where your body parts are, proprioception, is your brain's model of body configuration. That model can be wrong. In experiments where people watch a rubber hand being touched while their real hand is touched simultaneously, they start to feel like the rubber hand is their hand. The simulation has incorporated the fake hand into its body model.

This extends beyond immediate sensation. Your memory is simulation. When you remember something, your brain reconstructs a model of the past event using stored fragments. The reconstruction feels like playback, but it's actually generation. Your brain is simulating the past, creating a model that feels like memory but is really imagination constrained by stored data.

Your sense of time is simulation. Physical reality doesn't have past, present, and future in the way you experience them. Physics is reversible at fundamental levels. But your consciousness simulates time as flowing forward, as having an arrow, because that's useful for predicting and planning. The simulation creates the sense of temporal passage.

Even your sense of self is simulation. Your brain constructs a model of you, a self-representation that feels like the real you but is actually the brain's best guess about what kind of agent it's supposed to be controlling. This model includes your traits, your history, your goals, your boundaries. It's a working fiction, useful but not literally true.

So what you think of as consciousness is actually your brain running a simulation of yourself in a simulated world. The simulation is continuous, updating many times per second, incorporating new sensory data and predictions. It feels real because it's the only reality you have access to. You can't step outside the simulation to compare it with actual reality. The simulation is your reality.

This explains a lot. It explains why consciousness feels unified even though processing is distributed. The simulation presents a unified model. It explains why consciousness seems separate from the physical brain. You're experiencing the simulation, not the substrate running it. It explains why consciousness can be modified by beliefs and expectations. The simulation incorporates predictions and assumptions, so changing those changes the simulation.

It also explains the hard problem differently. You're not asking why matter produces consciousness. You're asking why the brain produces simulations, and why there's something it's like to be a simulation. These are still hard questions, but they're more tractable. Simulations are computational structures. Understanding what kinds of computational structures generate experience is difficult, but it's the right kind of question for science to address.

If consciousness is simulation, this changes how you think about reality. You're not in direct contact with the world. You're in contact with your brain's model of the world. The model is usually accurate enough for survival, but it's not perfect. It can't be. Perfect accuracy would require infinite processing power and infinite sensory data. The brain has to cut corners, make guesses, simplify.

This means your reality is necessarily distorted. Not hugely, not usually in ways that matter for everyday function. But systematically biased in predictable ways. Your brain's simulation has built-in assumptions that shape everything you experience.

For example, you see straight lines as straight and right angles as right angles because your visual system evolved to detect these features. But in physics, perfectly straight lines and perfect right angles might not exist. They're idealizations, simplifications your brain uses to make sense of messy reality. You see them everywhere because your simulation constructs them, not because they're really there.

Same with objects. You perceive the world as made of distinct separate things. This chair, that table, this person. But at the atomic level, there are no clean boundaries. Objects are clouds of particles loosely organized, constantly exchanging particles with their environment. The sharp boundaries are in your simulation, not in reality. Your brain constructs objects because treating reality as made of separate things is useful for acting in the world.

Even causation might be simulation artifact. You see events as causing other events, billiard balls colliding, switches turning on lights, your decisions causing your actions. But physics doesn't really have causation in the way you experience it. It has mathematical relationships between states. Your brain constructs the sense of causation because it's useful for prediction and control. The flow from cause to effect is in the simulation.

This doesn't mean nothing is real. The underlying reality exists. But your experience of it is heavily mediated by simulation. You're living in a model, a representation, not in direct contact with the thing itself. And the model is optimized for function, not for truth.

Understanding this changes your relationship with experience. You can start to notice where the simulation is making guesses, where it's filling in details, where it's constructing rather than receiving. You can catch it in the act of simulation.

This is partly what meditation does. By paying close attention to experience, you start seeing the seams in the simulation. You notice how visual detail disappears outside the focus of attention. How thoughts arise from nowhere without a thinker. How the sense of continuous self is actually reconstructed moment by moment. The simulation becomes visible as simulation.

Psychedelics do something similar but more dramatically. They disrupt normal simulation processes, letting you see how much of ordinary experience is constructed rather than given. When the simulation breaks down or runs in unusual modes, you realize how much of what seemed real was actually your brain's creative interpretation.

But you don't need drugs or meditation to notice this. Just pay attention. Notice how your expectations shape what you perceive. How your mood colors your interpretation of events. How your sense of time speeds up when you're engaged and slows down when you're bored. None of this is in the external reality. It's all in your simulation.

The practical implication is that changing your simulation changes your experienced reality. Not the underlying whatever-it-is, but reality as you experience it, which is the only reality you have access to. Change your predictions and your perception changes. Change your attention and what's real for you changes. Change your self-model and who you are changes.

This is why cognitive therapy works. It helps you notice that your simulation is running on bad assumptions. You're predicting negative outcomes, interpreting ambiguous events negatively, constructing a harsh reality from neutral inputs. Change the assumptions and the simulation changes. The same inputs generate different reality.

This is why placebo effects are real. Your brain's simulation includes expectations about how treatments work. When those expectations are activated, the simulation changes. Pain decreases, healing accelerates, symptoms improve. Not through imagination or willpower, but because the simulation genuinely runs differently when different predictions are loaded.

This is why beliefs about what's possible shape outcomes. If your simulation includes the assumption that you can't do something, it generates reality consistent with that assumption. You don't try, or you try half-heartedly, or you interpret setbacks as confirmation. The simulation creates self-fulfilling prophecies because it's not just observing reality, it's constructing it.

But there are limits. Your simulation can't create just anything. It has to stay roughly consistent with sensory input or you get hallucination and delusion. The simulation needs to track reality well enough for you to function. Too much divergence and you run into walls, literally and figuratively.

The skill is learning to modify your simulation in useful directions while keeping it calibrated to reality. Make it more flexible where rigidity causes problems. More accurate where accuracy matters. More positive where negativity doesn't help. But always keeping enough contact with ground truth that you can navigate effectively.

This is the art of consciousness modification we've been discussing throughout this book. You're not changing yourself directly. You're changing the simulation, the model your brain runs to generate your experience. And since the simulation is your conscious experience, changing it changes you.

If consciousness is simulation, then you are a simulation. Not in someone else's computer, but in your own brain. A model that the brain builds and maintains to help it function. The model includes everything you experience, including your sense of being something experiencing it.

This might sound diminishing. You're just a simulation? Not real, just a model? But the simulation is perfectly real as a simulation. It's real computation happening in real time on real hardware generating real outputs. The fact that it's simulation doesn't make it less real. It just means it's real in a different way than you thought.

And being simulation gives you leverage. Simulations can be modified. Parameters can be adjusted. Code can be rewritten. If you were a fixed thing, a permanent soul or unchangeable essence, you'd be stuck. But as simulation, you're fluid. You're whatever the current code generates. Change the code and you change who you are.

This is both liberating and vertiginous. Liberating because nothing about you is fixed. You're not trapped by your past, your genes, your circumstances. You're a real-time construction that can be reconstructed differently. Vertiginous because there's no solid ground. You're process, not substance. Pattern, not thing. Simulation, not reality.

But simulation is real. It's just real in the way software is real, not in the way hardware is real. Software is pattern, information, organization. But pattern and information and organization are perfectly real categories. They cause things to happen. They have effects in the world. Your consciousness, as simulation, is as real as anything else, just understood correctly as process rather than object.

And understanding yourself as simulation gives you the tools to work with yourself effectively. You stop trying to change the unchangeable essence and start working with the actual mechanics of how you're generated. You stop asking "what am I" and start asking "how am I implemented" and "can I run differently."

These are programming questions, not philosophical ones. And they have programming answers. You can debug your simulation. You can patch bugs. You can optimize performance. You can install new features. You can run in different modes for different contexts.

You're not just the user of your consciousness. You're also the developer. The code is running, but you can modify it while it runs. You're self-modifying code simulating a self that can modify its simulation. It's recursive and strange, but it's also practical.

Because if you understand how the simulation works, you can work with it rather than against it. You can notice its assumptions and biases. You can see where it's constructing rather than observing. You can catch it making predictions and update those predictions when they're wrong. You can reprogram your experience by reprogramming how your brain generates experience.

This is what we've been building toward all along. Consciousness as programming isn't just a theory. It's a practical framework for working with your own mind. Understanding that you're simulation gives you handles for modification. Understanding how the simulation is constructed tells you where to intervene. Understanding that the simulation is running in real time tells you that changes are always possible.

You're not stuck being the simulation you currently are. You're code that can rewrite itself, simulation that can redesign its own parameters, model that can update based on what it learns about being a model.

That's the central insight of thinking about consciousness as programming. You're not a thing that has consciousness. You're consciousness itself, understood as process rather than substance. And processes can change. Code can be rewritten. Simulations can run differently.

The question is what simulation you want to run. What reality you want to construct from the inputs you receive. What version of yourself you want to generate. These are choices, not fixed facts. Constrained choices, yes. You can't simulate anything you want. But within real constraints, there's genuine flexibility.

And using that flexibility wisely requires understanding what you actually are. Not a soul. Not a thing. Not even really a person in the traditional sense. A simulation. A running process. A pattern that maintains itself by continuously reconstructing itself.

That's what consciousness is if consciousness is programming. And if that's what you are, then understanding yourself means understanding the code. And understanding the code means you can modify it.

Which is where we're going in the final chapters. Not just what the code is, but how to work with it, how to debug it, how to optimize it, how to rewrite it when necessary. How to be a good programmer of your own consciousness.

Chapter Eleven

DEBUGGING YOUR CODE

You have bugs. Everyone does. Patterns in your thinking and behavior that don't work, that create problems, that keep running even though they make things worse. These are bugs in your consciousness code. And just like software bugs, they can be identified, understood, and fixed.

The first step is recognizing you have bugs at all. Most people live with their bugs for years without noticing them. The bugs feel normal because they've been running so long. You think that's just how you are. But it's not. It's how your current code runs. Different code would run differently.

Common bugs include loops that never exit. You think the same thoughts repeatedly without resolution. You worry about the same things in the same ways. You have the same arguments with the same people about the same issues. The code keeps executing the same routine, getting the same results, but never breaking out to try something different.

Another common bug is faulty error handling. Something goes wrong, and instead of dealing with it, your code crashes or freezes. You get overwhelmed, shut down, can't think clearly. Or you panic and make things worse. The bug is in how your system responds to errors, not in having errors at all.

Then there's the bug where your code makes assumptions that used to be true but aren't anymore. You learned patterns as a child that made sense then. But you're not a child anymore. The patterns still run though, generating responses that don't fit your current life. Old code that should have been updated but wasn't.

Memory leaks are another problem. You're supposed to process experiences and let them go. Instead, certain memories keep using resources, staying active, draining energy. Traumatic events that should have been stored and moved on from instead keep replaying, consuming processing power that should be available for current tasks.

There are also conflicts between different parts of your code. One subroutine wants safety. Another wants excitement. They fight for control, creating internal conflict and inconsistent behavior. Neither runs properly because they're interfering with each other. The system needs integration, but the parts are competing.

Recognizing bugs requires observation. You need to watch your own processing, notice when things go wrong, identify the patterns. This is harder than it sounds because your bugs are part of your observing system. You're using buggy code to examine buggy code. But it's possible with practice.

Start with outcomes. When things go badly, work backward. What were you thinking? What were you feeling? What pattern was running? Don't judge it as good or bad yet. Just identify it. This thought led to that feeling led to this action led to that outcome. Map the chain.

Do this repeatedly and patterns emerge. The same thought patterns precede the same problems. The same triggers produce the same responses. The same situations activate the same bugs. Once you can see the pattern, you've identified a bug.

Now comes debugging. The standard programming approach is to understand what the buggy code is trying to do, figure out why it's not working, and rewrite it to work correctly. Same process applies to consciousness bugs.

What is the pattern trying to accomplish? Usually something sensible, at least originally. Worry tries to prevent bad outcomes. Anger tries to protect boundaries. Avoidance tries to prevent pain. The pattern had a purpose. It's just not accomplishing that purpose effectively anymore.

Why isn't it working? Often because the situation has changed but the pattern hasn't updated. The old solution doesn't fit the new problem. Or the pattern was never good at its intended purpose, just the best your system could come up with at the time. Or the pattern creates side effects worse than the problem it's solving.

Understanding this helps. You stop seeing the bug as enemy and start seeing it as failed attempt at solution. This makes it easier to work with instead of fighting against. You're not eliminating a bad part of yourself. You're updating code that's trying to help but doing it badly.

The rewrite starts with noticing when the bug runs. You can't change it if you don't catch it. So you practice awareness. When you feel the familiar pattern starting, you notice it. This is the bug. Right here. Running now.

At first, noticing comes too late. The pattern has already run, caused problems, and you realize afterward. That's fine. Noticing after is better than not noticing at all. Keep practicing and you start noticing during. The pattern is running but you catch it mid-execution. Eventually, you notice before. You feel the trigger, see the pattern about to start, and have a moment to choose.

That moment of choice is where the rewrite happens. You've caught the bug before it fully executes. Now you need an alternative. What do you run instead? This is where having analyzed the pattern helps. You know what it's trying to accomplish. You just need a better way to accomplish it.

The new pattern doesn't have to be perfect. It just has to be better than the bug. If worry is trying to prevent bad outcomes, maybe planning is a better approach. Actually think through what could go wrong and what you'd do, then let it go. That accomplishes the same goal with less repeated mental spinning.

If anger is trying to protect boundaries, maybe clear communication works better. State what you need instead of exploding. That protects boundaries without the side effects of anger damaging relationships. Same purpose, better implementation.

If avoidance is trying to prevent pain, maybe gradual exposure is better. Face the thing in small doses, build tolerance, discover it's not as bad as feared. That reduces pain long-term better than avoidance which just postpones and amplifies it.

The key is having the alternative ready before the moment arrives. You can't invent new patterns under pressure. The bug runs too fast. You need to prepare the patch ahead of time. Practice the new response when you're calm. Visualize it. Walk through it mentally. Make it familiar so it's available when needed.

Then, when the moment comes and you've noticed the bug starting, you run the patch instead. Deliberately. Consciously. It feels awkward at first. The old pattern is automatic, smooth, familiar. The new one is effortful, clunky, strange. That's normal. New code always feels awkward before it becomes automatic.

But if you keep running the new pattern, it gradually becomes automatic too. The brain rewires. The new connections strengthen. What was effortful becomes easy. Eventually, the new pattern becomes the default and the old bug stops running entirely. You've successfully patched your code.

This takes time. Bugs that have been running for years don't get fixed in a day. You'll run the old pattern sometimes out of habit. You'll forget to use the patch when stressed. You'll partially execute both patterns and create new confusion. That's part of debugging. You iterate. You keep refining.

Some bugs are harder to fix than others. Deep patterns, established early, wired into fundamental assumptions, these resist change. You might need to go deeper than just replacing one behavior with another. You might need to examine the underlying beliefs and assumptions the bug is built on.

For example, if you have a pattern of self-criticism, the surface bug is the critical thoughts. But underneath might be a belief that you need to be perfect to be acceptable. That belief generates the criticism as a quality control mechanism. Patching just the critical thoughts won't work long-term if the perfectionism belief remains. You need to debug the deeper layer.

This is where therapy helps. A good therapist helps you identify bugs you can't see yourself, trace them to their sources, understand why they run, and develop patches. They provide external observation of your code that you can't get from inside. And they help you test patches before deploying them, working through obstacles in a safe space.

But you can do significant debugging on your own if you're willing to observe yourself honestly and experiment with changes. The key is persistence. Don't expect quick fixes. Complex code takes time to refactor. Small improvements compound. Keep at it.

There's also preventive debugging. Instead of waiting for bugs to cause problems, you can audit your code proactively. Look for patterns that might become bugs before they do. Notice where your thinking is rigid, where you're making untested assumptions, where you're avoiding looking at something. These are potential bugs developing.

Regular practices help with this. Journaling lets you observe your thought patterns. Meditation lets you notice mental habits. Therapy provides regular checkups. Time with people who know you well and will give honest feedback reveals blind spots. These are all ways to catch bugs before they cause major problems.

You can also work on making your code more robust in general. Build in better error handling so problems don't crash the system. Create flexibility so you can adapt when situations change. Develop awareness that catches bugs early. These improvements make all your patterns run better, not just fixing specific bugs.

Think of it like upgrading your operating system. Individual bug fixes help, but sometimes you need deeper changes to how the whole system works. Better memory management. More efficient processing. Improved integration between components. These system-level improvements make everything work better.

But you have to be careful. Aggressive debugging can break things. If you try to change too much too fast, you can destabilize your whole system. You need to change incrementally, test thoroughly, make sure new code actually works before deploying it everywhere. Slow and steady beats rushing.

Also, some apparent bugs might not be bugs. Some patterns that look dysfunctional might be working as designed for purposes you don't understand. Before you remove something, make sure you know what it's doing. You might discover it's solving a problem you didn't know you had. Take it out and the problem appears.

This is why self-compassion matters during debugging. You're working with code you didn't write originally. Evolution wrote some. Childhood experience wrote some. Random life events wrote some. You inherited this code, and some of it is messy. But it kept you alive and functioning this long. It deserves some respect even when it needs updating.

The goal isn't to eliminate all bugs and run perfectly. That's impossible. The goal is to identify the bugs causing real problems, understand them, develop better alternatives, and gradually improve how your code runs. You'll always have some bugs. That's part of being conscious. But you can reduce them, minimize their impact, and keep getting better at debugging as you go.

And the process of debugging changes you. Not just by removing bugs, but by developing the capacity to observe your own code and modify it. This meta-skill, being able to work on yourself systematically, is more valuable than any individual bug fix. Once you know how to debug, you can keep improving indefinitely.

You become your own developer. Not just running the code you were given, but actively maintaining and improving it. Identifying problems, researching solutions, implementing changes, testing results, iterating. This is consciousness taking responsibility for itself, using its self-modifying nature deliberately instead of letting changes happen randomly.

This is what separates humans from other conscious systems. Other animals are conscious, probably, but they can't debug themselves this way. They're stuck running their code as-is. Humans can examine their code, understand it, and rewrite it. This capacity for deliberate self-modification might be the most important feature of human consciousness.

And it's a skill that improves with practice. The more you debug, the better you get at it. You learn to recognize patterns faster. You develop better patches. You understand your own architecture. You become fluent in your own code. What seemed impossible at first becomes routine.

This doesn't mean the process ever ends. You'll find new bugs. You'll discover deeper patterns underneath patterns you already fixed. You'll face new situations that expose weaknesses in code that was working fine before. Debugging is ongoing maintenance, not a one-time project.

But that's okay. Actually, it's good. It means you're never stuck. There's always room for improvement. Always possible to run better. Always new versions of yourself you can become through careful code modification.

The question is whether you'll do it deliberately or let it happen randomly. Your code will change either way. Experience modifies you constantly. But conscious debugging lets you guide the changes toward improvement instead of just accepting whatever random modifications life throws at you.

You're the programmer. You're also the program. But you have some choice in what the code becomes. Not unlimited choice. You're constrained by your hardware, your history, your circumstances. But within those constraints, real choices exist. Real improvements are possible.

Debugging yourself is how you exercise that choice. How you take your self-modifying nature and point it in useful directions. How you become, gradually, the version of yourself you want to be rather than just the version you happened to become.

This is consciousness as programming made practical. Not just understanding what you are, but using that understanding to become something better. Code that debugs itself, improves itself, optimizes itself over time.

That's what we'll explore next. Not just fixing bugs, but actually optimizing the code to run better, faster, more efficiently. Taking consciousness from barely working to working well.

Chapter Twelve

OPTIMIZING CONSCIOUSNESS

Debugging fixes what's broken. Optimization makes what's working work better. Once you've handled the major bugs, you can focus on improving performance. Making your consciousness run faster, smoother, more efficiently. Using less energy for the same results or getting better results from the same energy.

The first area to optimize is attention. Attention is your most limited resource. You can only pay attention to a few things at once. Where you direct attention determines what enters consciousness, what gets processed deeply, what you remember. Most people use attention inefficiently, letting it scatter randomly or get captured by whatever's most stimulating.

Optimized attention means controlling where it goes. You decide what's important and point attention there deliberately. You resist distraction, notice when attention wanders, bring it back without fighting. This sounds simple but it's hard. Your attention has been trained by algorithms and advertisers to jump around constantly, to seek novelty, to follow interruption.

Training attention requires practice. Meditation is basically attention training. You pick a target, breath or body sensation or sound, and practice keeping attention there. When it wanders, you notice and return. Over and over. This builds the mental muscle of voluntary attention control.

With trained attention, everything improves. You think more clearly because attention stays on the problem instead of drifting. You remember better because attention creates stronger encoding. You work more efficiently because you're not constantly switching. You notice more because you're actually looking instead of being on autopilot.

The next optimization target is energy management. Consciousness uses a lot of energy. Your brain is two percent of your body weight but uses twenty percent of your energy. Running consciousness poorly wastes this expensive resource. Running it well conserves energy for when you actually need it.

Common energy waste includes unnecessary worry. Thinking about problems you can't solve right now. Replaying past events you can't change. Running mental simulations of feared futures. All this processing burns energy without producing useful results. It's like leaving programs running in the background eating up processor power for no reason.

Optimized energy management means shutting down unnecessary processing. You think about problems when you can actually do something about them. Otherwise, you let them go, conserve the energy. This isn't avoiding responsibility. It's being smart about when to deploy your limited resources.

Sleep optimization is crucial here. Sleep is when consciousness does maintenance, consolidates memories, clears metabolic waste, repairs damage. Skimp on sleep and everything runs worse. You're basically trying to run intensive programs on insufficient power. The code might execute but it runs slowly, makes mistakes, crashes under load.

Good sleep hygiene means regular schedule, dark quiet environment, wind-down time, no screens before bed. These aren't luxuries. They're necessary system maintenance. Ignore them and your consciousness runs at reduced capacity. Respect them and you operate at full power.

Another optimization is reducing context switching. Every time you switch tasks, there's overhead. You have to load the new context, remember where you were, figure out what to do next. Switch too often and you spend all your energy on switching instead of actually accomplishing anything.

Batch similar tasks together. Do all your emails at once instead of scattered through the day. Have deep work blocks where you focus on one thing for hours. Minimize interruptions. Turn off notifications. Close unnecessary tabs. Create conditions where your consciousness can load a context and stay in it long enough to actually get something done.

Decision fatigue is another energy drain. Every decision uses willpower, and willpower is limited. Make too many decisions and your decision-making quality drops. You start taking shortcuts, defaulting to easy choices, avoiding decisions entirely. This is why successful people often standardize small decisions, wearing the same outfit, eating the same breakfast. They're conserving decision energy for choices that actually matter.

You can do the same. Identify which decisions actually need careful thought and which can be automated or randomized. Save your decision-making capacity for high-stakes choices. For low-stakes stuff, just pick something and move on. The mental energy saved is worth more than getting the optimal choice on trivial matters.

Emotional regulation is another key optimization. Emotions are useful signals but they can also hijack your system, forcing everything else to stop while you deal with the emotional content. Unregulated emotions create constant interruptions, draining energy and preventing focus.

Optimized emotional processing means feeling emotions without being overwhelmed by them. You notice the feeling, understand what it's signaling, respond appropriately, then let it pass. The emotion gets acknowledged and processed instead of suppressed, but it doesn't take over your whole system.

This is different from suppressing emotion, which actually uses more energy long-term. Suppression requires constant effort to keep the emotion pushed down. Processing and release lets the emotion complete its cycle naturally. The first creates chronic drain. The second handles the emotion efficiently and moves on.

Memory optimization is another area with huge payoff. Your memory system is constantly active, storing, retrieving, consolidating. Run it inefficiently and you waste energy remembering useless stuff while forgetting important things. Optimize it and you keep what matters accessible while letting go of clutter.

The key is distinguishing between what needs to be in active memory versus what can be stored externally. You don't need to remember everything. You need to remember where to find things and how to use them when needed. This is why writing things down works. You're offloading storage to external systems, freeing up internal resources.

Note-taking, lists, calendars, reminders, these are all memory optimization tools. They let you stop using mental energy trying to remember stuff and trust your external systems instead. Your consciousness can focus on thinking and deciding instead of constantly trying not to forget things.

The same goes for skills. Practice makes things automatic, moving them from conscious execution to unconscious execution. This frees up consciousness for new learning. Early in learning anything, you have to think about every step. But with enough practice, the skill runs automatically and consciousness barely needs to be involved.

This is optimization through automation. Move routine operations to automatic processing so consciousness can handle exceptions and new situations. Your limited conscious capacity goes to things that actually need it instead of being wasted on stuff that could run automatically.

Physical optimization matters too. Your consciousness runs on biological hardware. The hardware's condition affects performance. Poor nutrition, dehydration, lack of exercise, chronic pain, all these degrade how well your consciousness operates.

You probably notice you think better after exercise, sleep, and good food. That's not coincidence. The physical state directly affects cognitive function. Optimize the hardware and the software runs better. Neglect the hardware and even optimal code performs poorly.

This means taking care of your body isn't separate from optimizing consciousness. They're the same project. The brain is an organ that needs proper conditions to function well. Provide those conditions and consciousness runs smoothly. Deprive them and everything struggles.

Social optimization is less obvious but equally important. Humans are social creatures. Your consciousness evolved to function in social contexts. Isolation degrades performance. Good relationships improve it. This isn't just about happiness. It's about how well your system works.

Social connection provides cognitive benefits. Explaining things to others clarifies your own thinking. Hearing different perspectives reveals blind spots. Collaboration distributes cognitive load. Emotional support reduces stress that impairs thinking. Social optimization means deliberately building and maintaining relationships that support cognitive function.

This doesn't mean maximum social contact. It means right-sized social contact with the right people. Some people need lots of interaction. Others need very little. Some people benefit from deep conversations. Others from casual companionship. Optimize for what actually helps your consciousness run well, not what you think you should do socially.

Environment optimization is another leverage point. Your physical environment affects consciousness more than you realize. Cluttered space creates cluttered thinking. Noisy environments fragment attention. Bad lighting strains processing. Uncomfortable conditions create constant low-level stress.

Optimize your environment and consciousness runs smoother. Clean organized space makes it easier to focus. Good lighting reduces strain. Comfortable temperature removes distraction. You don't need perfection. You need good enough that the environment supports rather than hinders your cognitive function.

Information diet optimization matters too. You consume information constantly. News, social media, messages, content. Most of it is junk food for consciousness. It tastes good in the moment but provides no nutrition and creates problems long-term.

Optimize by being selective about what information you consume. High-quality sources over random feeds. Books over tweets.

Long-form over snippets. Information that actually helps you think better rather than just consuming your attention. This isn't about being elitist. It's about feeding your consciousness input that makes it run well instead of junk that clogs it up.

All these optimizations compound. Better attention lets you sleep better. Better sleep improves emotional regulation. Better emotional regulation improves relationships. Better relationships reduce stress. Reduced stress improves decision-making. Better decisions improve outcomes. Better outcomes reduce anxiety. Less anxiety improves attention. The positive feedback loops build on each other.

This is why small optimizations matter. Each one creates marginal improvement. But marginal improvements compound. Over time, your consciousness runs significantly better than it did before, not from any single change but from many small optimizations accumulating.

The process is iterative. You optimize one thing, that reveals the next bottleneck, you optimize that, which reveals another. You never reach perfect optimization. There's always room for improvement. But each iteration makes things better.

And the improvements are real. This isn't just subjective feeling better. Optimized consciousness actually performs better. You think more clearly, remember more accurately, decide more wisely, act more effectively. The measurable outputs improve.

This is why optimization is worth the effort. You're not just making yourself feel better. You're making yourself work better. Your consciousness becomes a more capable tool for whatever you're trying to accomplish in life.

The key is approaching it systematically. Don't try to optimize everything at once. Pick one area, work on it until improvement becomes automatic, then move to the next. Build up optimization gradually, let each change stabilize before adding more.

And measure results. Notice what actually makes things better versus what you think should help but doesn't. Your system might not respond to optimizations the way the general advice suggests. What works for most people might not work for you. What shouldn't matter might matter a lot for your particular setup.

This is why you need to treat yourself as an experiment. Try optimizations, observe results, keep what works, discard what doesn't. You're debugging and optimizing code that's unique to you. Generic advice helps but needs to be customized to your specific implementation.

The goal isn't to be perfect. It's to be better. To run more smoothly. To waste less energy. To accomplish more with the same resources. To reduce friction. To make consciousness feel less like struggle and more like flow.

Optimized consciousness isn't necessarily easier. You might actually work harder because you can. But it's less effortful. Things that were difficult become natural. What required forcing becomes automatic. You're working with your system rather than fighting it.

This is the difference between raw intelligence and optimized consciousness. Intelligence is your hardware capability. Optimization is how well you use that capability. You can be very intelligent but run poorly, wasting your potential on inefficiency. Or you can be average intelligence but optimized, getting far more from what you have.

And optimization is more accessible than intelligence enhancement. You can't easily upgrade your hardware. But you can definitely improve how you use it. The returns are immediate and practical. You don't need to be smarter. You need to be better at being as smart as you already are.

This is consciousness as programming made practical. You're not just understanding the code or fixing bugs. You're tuning the system for performance. Making it faster, smoother, more reliable. Getting better results from the same hardware through better software configuration.

The remaining chapters will take this further. Not just optimizing what you have but potentially adding new capabilities. Not just running better but running in new ways. Not just improving current consciousness but exploring what consciousness could become.

But optimization is the foundation. Get good at running what you have before trying to build something new. Make your current code efficient before adding features. Master the basics before attempting advanced modifications.

Because optimized consciousness isn't just better for you. It's the platform for everything else. All the transformations we've discussed, debugging, reprogramming, consciousness design, they all work better when running on optimized systems. Optimization multiplies everything else.

So start there. Pick one thing. Optimize it. Move to the next. Build up gradually. Let the improvements compound. And watch as your consciousness transforms from struggling to running smoothly, from fighting itself to working with itself, from barely functional to genuinely optimized.

That's when things get interesting. That's when new possibilities open up. That's when consciousness stops being a problem to solve and starts being a tool you can actually use effectively.

Chapter Thirteen

INSTALLING NEW MODULES

Once your code is debugged and optimized, you can start adding new capabilities. Skills, perspectives, ways of thinking that weren't there before. You're not just fixing or improving what exists. You're installing completely new modules that expand what your consciousness can do.

Think of learning a language. Before you learn it, you can't think in that language. The concepts it contains, the ways it structures reality, these aren't accessible to you. After you learn it, you have new ways of thinking. Some ideas exist more naturally in certain languages. Learning the language installs those ideas as available thought patterns.

Or consider learning math. Before calculus, you can't think about rates of change in sophisticated ways. After calculus, you have tools for understanding dynamic systems. The math is a module that gives your consciousness new processing capabilities. Problems that were unsolvable become solvable because you installed the right tools.

This is true for any skill. Learning to draw installs visual processing modules. Learning music installs auditory pattern recognition. Learning to code installs logical structures and debugging habits. Each skill is new software that extends what your consciousness can do.

But you can also install more abstract modules. Mental models, frameworks, perspectives. These change how you process information even when you're not consciously using them. They become part of your background processing, shaping how you interpret everything.

For example, learning about cognitive biases installs awareness of how your thinking can go wrong. You start noticing when you're making biased judgments. The knowledge becomes a filter that catches errors you wouldn't have seen before. It's not just information. It's a new processing layer.

Or learning about systems thinking changes how you see causation. Instead of simple cause and effect, you start seeing feedback loops, emergent properties, unintended consequences. The framework becomes part of how you automatically analyze situations. It's installed as permanent modification to your thinking apparatus.

Meditation installs an observer module. Before practice, you're identified with your thoughts and feelings. You are your mental content. After sustained practice, there's a part of consciousness that can observe mental content without being it. The observer module creates distance, space, choice. It's a new capability that wasn't there before.

Therapy can install emotional processing modules. You learn to recognize feelings, name them, understand what they're signaling, respond appropriately. These become automatic over time. You're not following learned steps consciously. The processing just happens. The therapy installed new code that runs on its own.

The key to successful module installation is practice. You can't just read about something and have it installed. You have to use it repeatedly until it becomes integrated. Reading about meditation doesn't install the observer module. Meditating daily for months does. Reading about cognitive biases doesn't install the error-checking. Actively catching your biases repeatedly does.

This is the difference between knowing and knowing how. Knowing is information. Knowing how is installed capability. You can know lots of things without being able to do them. The doing requires installation through practice, turning information into functioning code.

Some modules are easier to install than others. Simple skills with clear practice paths install relatively quickly. Complex skills with fuzzy boundaries take longer. Abstract frameworks that require changing basic assumptions take even longer. The deeper the installation needs to go, the more practice required.

There's also interference between modules. Sometimes new modules conflict with existing code. Learning to think one way makes it harder to think another way. This is why experts can be blind to simple solutions. Their expert modules override basic approaches. The sophisticated code blocks simpler but sometimes better alternatives.

This means you need to be strategic about what modules you install. More isn't always better. Each module uses resources, creates overhead, changes how the whole system runs. Install too many and they interfere with each other. Install the wrong ones and they create problems despite being individually useful.

The question is what modules actually help. This depends on your goals, your life, your existing capabilities. Someone who works with people needs different modules than someone who works with abstract systems. An artist needs different modules than an engineer. There's no universal optimal module set.

But some modules are broadly useful. Mental models for understanding complex systems. Emotional awareness for navigating social reality. Attention control for basically everything. Critical thinking to avoid being manipulated. These have wide application regardless of specific goals.

Other modules are specialized. You install them because they're needed for particular purposes. Musical training if you want to make music. Statistical thinking if you work with data. Legal frameworks if you practice law. These are tool modules, useful when you need them, not relevant otherwise.

The installation process has stages. First is awareness. You discover the module exists, understand what it does, see why it might be useful. This is learning about the capability. You know it's possible but you can't do it yet.

Second is awkward practice. You try to use the module deliberately. It's slow, effortful, error-prone. You have to think about every step. The code is there but not integrated. It runs only when you consciously execute it. This stage feels clumsy and frustrating but it's necessary.

Third is integration. With enough practice, the module starts running without conscious effort. You don't have to think about the steps anymore. The capability becomes automatic. It's fully installed, running in the background, available when needed without deliberate activation.

Fourth is mastery. The module runs smoothly, efficiently, reliably. It's fully integrated with your other capabilities. You can use it creatively, combine it with other modules, apply it in novel situations. It's become part of your natural cognitive repertoire.

Not every module needs to reach mastery. Some are useful at integration level. You have them available when needed but don't need to be expert. The question is what level of installation serves your purposes. Sometimes basic competence is enough. Sometimes you need deep mastery.

The time investment differs dramatically between levels. Getting to awkward practice might take days or weeks. Integration takes months to years. Mastery takes years to decades. You need to choose wisely what's worth pushing to higher levels because you don't have unlimited time.

There's also maintenance. Installed modules degrade without use. Skills rust. Frameworks fade. Languages are forgotten. The code is still there but gets weaker. You need periodic practice to keep modules functional. The more modules you have, the more maintenance required.

This creates practical limits on how many modules you can maintain. You can install many, but you can only keep so many active and well-maintained at once. The rest will degrade to dormant states where they're sort of there but not really functional without reactivation.

Strategic choices about modules become crucial. What capabilities do you actually need? What's worth maintaining versus letting fade? What new installations are worth the time investment? These aren't just practical questions. They're questions about who you want to be.

Because installed modules change you. They're not just tools you pick up and put down. They become part of how you think, how you see, how you process reality. Install enough modules in a particular domain and you become a different kind of person. An expert, a specialist, someone whose consciousness operates differently than it did before.

This is both powerful and dangerous. Powerful because you can deliberately develop capabilities, become able to do things you couldn't do before. Dangerous because you might install modules that take you in directions you later regret, or that make you less able to do other things you value.

There's also the question of what modules are even available. You can only install what you know exists. Most people go through life using the default modules they happened to acquire through education and experience. They don't know what other options exist.

Exploring consciousness possibilities means discovering what modules are installable. What frameworks exist? What skills are learnable? What perspectives are accessible? The space of possible modules is huge and mostly unexplored by any individual.

This is where learning about different traditions helps. Philosophy, religion, art, science, different cultures, different historical periods, all have developed distinct modules. Ways of thinking and being that aren't obvious from your starting point. Exposure to diverse approaches reveals possibilities you didn't know existed.

Sometimes the most valuable module installations come from completely unexpected sources. A random book, a chance conversation, an unusual experience that shows you a way of being you never considered. Staying open to surprise installations can be more valuable than systematically pursuing known modules.

But you also need discernment. Not every possible module is worth installing. Some are harmful. Some are useless. Some seem valuable but create more problems than they solve. You need to evaluate before investing installation time.

Signs of valuable modules include broad applicability, compatibility with existing code, clear improvement in capability, and lasting value. Signs of problematic modules include narrow usefulness, conflicts with important existing capabilities, making things harder instead of easier, and quickly becoming obsolete.

You can test modules before full installation. Try them temporarily, see how they work, decide if they're worth keeping. Don't commit to full integration until you've verified the module actually helps. This is like running software in trial mode before purchasing.

Some people are collectors, installing many modules superficially without deep integration. Others are specialists, installing few modules but driving them to mastery. Both approaches have merits. The collector has flexibility and breadth. The specialist has depth and reliability. Most people fall somewhere in between.

But there's a danger in staying at beginner level with everything. Partially installed modules that never integrate don't actually add capability. They're just information taking up space. Better to fully install fewer modules than partially install many.

This is why focus matters. Pick modules that align with your goals and values. Install them properly. Get them to integration or mastery. Then consider what to add next. Sequential installation with full integration beats parallel installation with shallow learning.

The ultimate goal is a module set that works together, that complements rather than conflicts, that gives you the capabilities you actually need for the life you want to live. This takes time to develop. You'll install things that turn out not to help. You'll discover needs for modules you didn't anticipate. The set evolves as you learn what actually matters.

And some modules open access to other modules. Meta-learning modules that make learning easier. Framework modules that organize other knowledge. Integration modules that help different capabilities work together. These multiplier modules are especially valuable because they improve everything else.

Learning how to learn is a meta-module. Once installed, it makes installing other modules faster and more effective. Critical thinking is a meta-module. It helps you evaluate what modules are worth installing. Self-awareness is a meta-module. It helps you notice what's working and what isn't.

Investing in meta-modules pays long-term dividends. They're harder to install because they're more abstract. But they multiply the value of everything else you learn. They're infrastructure that supports all other capabilities.

The vision here is consciousness as customizable. You're not stuck with the modules you happened to acquire. You can deliberately add capabilities. You can become able to do things you currently can't do. You can expand what your consciousness is capable of through systematic module installation.

This is programming yourself, adding features, extending functionality. Not changing what you fundamentally are, but expanding what you can do with what you are. Like upgrading software to add new features while keeping the basic system the same.

The question is what features you want. What capabilities would make your consciousness more powerful, more useful, more aligned with your goals? The answers are personal. There's no universal optimal module set. But thinking about the question seriously can guide your learning and development.

Because every skill you learn, every framework you adopt, every practice you maintain is installing modules. The only question is whether you're doing it deliberately or randomly. Whether you're choosing what to install or just accepting whatever comes along.

Consciousness as programming means you're the developer. You can choose what features to add, in what order, to what depth. The installation takes work. But the payoff is genuine expansion of capability, becoming able to do things you literally couldn't do before.

That's the power of module installation. Not just improving what you have, but adding what you don't have. Not just optimizing existing code, but writing new code that gives your consciousness new powers. Transforming what you're capable of through deliberate feature addition.

And this brings us to the final question. What's the limit? How far can consciousness be expanded? What's the maximum capability you could theoretically develop? Is there a ceiling or can consciousness keep growing indefinitely?

We'll explore that next. Not just what consciousness is or how it works, but what it could become with unlimited development. The far edge of possibility. The ultimate versions of consciousness as programming.

Chapter Fourteen

THE LIMITS OF CONSCIOUSNESS

Is there a ceiling? Can consciousness keep expanding indefinitely or does it hit walls? This isn't just philosophical. Understanding limits helps you aim efforts appropriately. No point trying to exceed impossible limits. But if limits are further out than you thought, you might be aiming too low.

Physical limits exist for sure. Your brain is finite. It has limited neurons, limited connections, limited processing speed, limited energy. You can optimize within these constraints but you can't exceed them. No amount of training will let you remember every moment of your life perfectly or think a thousand times faster than normal.

But the constraints might be less restrictive than they seem. Your brain isn't using anywhere near its theoretical capacity. Most estimates suggest we're using a small fraction of possible connections, possible patterns, possible configurations. There's enormous room for growth before hitting actual hardware limits.

Consider memory. You forget most of your life. This feels like a limit. But memory champions demonstrate that with the right techniques, memory capacity is far larger than normal use suggests. The limit isn't storage space. It's how you encode and retrieve. Better strategies dramatically expand what's possible.

Or attention. Normal people can focus on maybe one complex thing at a time. But experienced meditators report being able to hold multiple objects in awareness simultaneously without losing clarity on any of them. The attention system can be trained beyond normal bounds. The limit isn't fixed. It's expandable.

Processing speed feels fixed. You think as fast as you think. But research on expertise shows experts process domain-specific information faster than beginners. Chess masters see patterns instantly that take amateurs minutes to notice. The speed-up comes from pattern recognition, chunking information into larger units. Better organization leads to faster processing.

So some apparent limits are really just default settings. Train the system and it performs beyond what seemed possible. The ceiling is higher than it appears from untrained baseline.

But there are probably real limits too. Information processing speed is constrained by neural firing rates. Memory is constrained by total available synaptic connections. Attention is constrained by the architecture that coordinates different brain regions. These are hardware limits that training can't overcome, only approach more closely.

The question is where the real limits are versus where we assume limits are. Most people hit assumed limits, barriers they think exist, long before approaching real limits. They stop trying to improve because they've reached what they think is possible. But often, pushing past assumed limits reveals more room than expected.

This is why mindset matters. If you think your capabilities are fixed, you'll stop at the first difficulty. If you think capabilities are expandable, you'll keep pushing. The belief itself influences whether you reach actual limits or just assumed ones.

Cultural limits matter too. Your culture teaches you what's normal, what's possible, what's worth attempting. These cultural assumptions create invisible boundaries. Most people never test them. They accept the cultural default as natural limit.

But looking across cultures reveals huge variation in what's considered normal. Meditation capabilities developed in Buddhist cultures seem superhuman from Western baseline. Mathematical abilities developed in certain educational systems seem extraordinary to others. The variations suggest cultural defaults are nowhere near biological limits.

Even within cultures, there are outliers. People who exceed normal bounds in specific domains. Athletes, artists, scholars, meditators, savants. They prove that standard limits aren't absolute. The question is what allows some people to exceed standard bounds while most stay within them.

Part of it is dedication. Outliers practice obsessively. Ten thousand hours is the commonly cited number for expertise, though the actual requirement varies by domain. But extended focused practice clearly pushes capabilities far beyond initial levels. Most people simply don't invest enough time to find their real limits.

Part of it is method. Not all practice is equal. Deliberate practice, focused on weak points, with immediate feedback, produces far better results than just doing something repeatedly. The people who exceed standard bounds usually aren't just working harder. They're working smarter, using better training methods.

Part of it might be genetic. Some variation in capability is innate. People start at different baselines, have different hardware. But even here, training effects are usually larger than genetic effects within normal ranges. The genetic component sets the starting point and maybe the ultimate ceiling, but there's huge room for development between them.

The limits question gets more interesting when you consider consciousness modification beyond normal training. What if you could change the hardware, not just optimize the software? What becomes possible then?

Drugs are one approach. They temporarily modify how the hardware runs. Stimulants speed up processing. Psychedelics change pattern recognition. Nootropics supposedly enhance various cognitive functions. The effects are temporary and often come with costs, but they demonstrate that consciousness capabilities aren't fixed by your standard brain chemistry.

Some people use drugs strategically for temporary enhancement. Students taking stimulants for exams. Programmers using substances to focus. Artists using psychedelics for creative breakthroughs. This is pharmacological consciousness modification, changing what you can do by changing brain chemistry.

But the approach has obvious problems. Side effects. Addiction risk. Tolerance. Long-term health impacts. Using drugs to exceed normal limits trades future capability for present performance. The costs eventually exceed the benefits for most people.

More interesting is whether you could achieve similar effects through training alone. Experienced meditators report states that sound similar to drug effects but without the drugs. Flow states produce focus comparable to stimulants. Deep meditation produces experiences comparable to psychedelics. The brain might be capable of generating enhanced states naturally if trained properly.

This suggests limits are partly about accessing capabilities that already exist. Your brain can operate in various modes. Normal waking consciousness is one mode. But there are others, flow, meditation states, hypnosis, various altered states. Each mode has different capabilities. Learning to access them deliberately expands what you can do without changing hardware.

Technology offers another path past limits. External tools that extend cognitive capability. Writing extends memory beyond what brains can hold. Calculators extend arithmetic beyond what minds can compute. Computers extend information processing far beyond biological rates. The tools don't change your brain but they change what you can accomplish.

This is the extended mind thesis, that cognition extends beyond the skull to include tools and environment. Your consciousness isn't just your brain. It's your brain plus notebook plus smartphone plus computer plus library plus network of people. The system as a whole has far greater capabilities than the brain alone.

If consciousness is the system not just the biological part, then limits expand dramatically. You're not limited by brain capacity if you can outsource to tools. You're not limited by individual memory if you can access collective knowledge. You're not limited by personal processing speed if you can use computers.

But there's a question about what counts as you in this picture. If your capabilities depend on external tools, are they your capabilities or the tools' capabilities? If you can only do something with your phone, can you actually do it or does your phone do it?

This matters for understanding consciousness limits. If we're measuring biological consciousness alone, tools don't extend limits. They just compensate for them. But if we're measuring effective capability, the system you-plus-tools is what matters. The biological limits become less relevant.

Future technology might dissolve this distinction. Brain-computer interfaces could integrate external processing directly into consciousness. Neural implants could expand memory or speed. Genetic modification could alter brain development. These aren't available yet, but they're conceivable. They would genuinely push past biological limits rather than working around them.

Some transhumanists envision radical enhancement, consciousness running on artificial substrates, freed from biological constraints entirely. Unlimited memory, super-fast processing, direct information access, capabilities far beyond human baseline. Whether this is achievable or even desirable is debated, but it represents thinking about consciousness limits as temporary rather than permanent.

But even without technology, there's the question of collective consciousness. Individual humans have limits. But groups of humans working together achieve things no individual could. The group has capabilities beyond any member. Is there a sense in which the group is conscious?

We touched on this earlier with panpsychism. If consciousness exists at multiple scales, maybe human groups have group consciousness. Not just metaphorically but literally. The group processes information in integrated ways. It makes decisions. It has goals. It responds to its environment. Maybe that's enough for it to be conscious at group level.

If so, then consciousness limits change dramatically. You as individual have individual limits. But you as part of groups have access to group-level capabilities. Science is group consciousness processing questions too complex for individuals. Markets are group consciousness allocating resources. Cultures are group consciousness transmitting accumulated knowledge.

This isn't about hive minds or losing individuality. It's about recognizing that consciousness already operates at multiple scales. You're individual consciousness participating in larger collective consciousness. The limits at one scale don't constrain the other scale.

The practical upshot is that consciousness limits are fuzzy. There are hard physical constraints. But within those constraints, there's enormous flexibility. Training, tools, methods, collective organization, all can push capabilities far beyond standard baseline. Most people never approach real limits because they stop at assumed limits.

So when thinking about what consciousness can become, the answer is probably much more than currently exists in you. Not infinite. Not unconstrained. But far more developed than most people achieve. The ceiling is high enough that reaching it isn't the practical concern for most purposes.

The more relevant question is direction rather than limits. Not how far can consciousness develop, but in what direction should it develop. What capabilities matter? What enhancements are worth pursuing? What modifications serve actual goals rather than just being impressive?

This is where the programming metaphor comes full circle. You're not trying to maximize all possible capabilities. You're trying to build consciousness that works well for your purposes. That serves your values. That lets you live the kind of life you want to live.

Some capabilities matter for nearly everyone. Clear thinking. Emotional balance. Good attention control. Reliable memory for important things. These are general-purpose capabilities that improve most lives. Working on these makes sense for almost anyone.

Other capabilities are specialized. Extreme mathematical ability matters if you're a mathematician. Perfect pitch matters if you're a musician. Photographic memory might be useful for some purposes but unnecessary for most. These are worth developing if they serve your specific goals but not as general enhancement.

And some capabilities might not be worth pursuing even if possible. Eliminating all emotional responses might make decision-making more logical but it would also make life hollow. Remembering everything perfectly might be overwhelming rather than helpful. More capability isn't always better if it comes with costs.

The art is choosing wisely. Understanding what's possible helps you aim appropriately. Understanding what's valuable helps you choose what to aim for. Understanding what's achievable with reasonable effort helps you prioritize.

Consciousness development is a lifelong project. You're never done. There's always room for improvement, new capabilities to develop, bugs to fix, optimizations to make. But it's a project worth pursuing because consciousness is the interface through which you experience everything else.

Better consciousness means better life, not because life circumstances change but because your experience of those circumstances improves. You process more clearly, respond more appropriately, suffer less from avoidable mistakes, enjoy more because you're actually present.

This is the ultimate implication of consciousness as programming. You're not fixed. You're not stuck. You're code that can be modified, debugged, optimized, extended. The modifications take effort. But they're possible. And they matter.

The question we've been circling throughout this book is whether you'll take conscious control of your own development or let it happen randomly. Your consciousness will change either way. Experience modifies you constantly. But you can guide that modification, direct it toward improvements, become deliberately better rather than just different.

That's what consciousness as programming offers. Not a complete theory of what consciousness is. We're far from having that. But a practical framework for working with your own consciousness. For understanding yourself as process rather than thing. For recognizing that you're self-modifying code with some control over the modifications.

The final chapter will bring this together. Not just what consciousness is or how it works, but what it means to be conscious in this particular way, to be self-modifying code that can understand itself as code. To be programming that programs itself.

This is the strange loop at the heart of consciousness. The code examining itself, modifying itself, becoming something new through the act of understanding what it is. And in that loop lies both the mystery and the possibility of consciousness.

CONCLUSION: THE STRANGE LOOP

We've come full circle, but we're not where we started. Understanding consciousness as programming doesn't solve the hard problem. It doesn't explain why there's something it's like to be you. But it gives us a way to work with consciousness that's practical, useful, and transformative.

The central insight is simple. You're not a thing that has consciousness. You're consciousness itself, understood as process rather than object. You're code executing in real time, generating experience moment by moment. And that code can modify itself.

This changes everything about how you relate to your own mind. You stop treating yourself as fixed and start treating yourself as fluid. You stop accepting patterns as inevitable and start seeing them as changeable. You stop being passive recipient of your experience and start being active participant in constructing it.

The hard problem remains. We still don't know why information processing creates experience. Why there's something it's like to be a pattern of activity in a brain. Why consciousness exists at all rather than everything just happening in the dark. These questions might not have answers we can access from inside consciousness.

But we don't need to solve the hard problem to work with consciousness effectively. We don't need to know why consciousness exists to understand how it works, how to improve it, how to modify it. The practical questions are answerable even if the philosophical ones remain open.

Throughout this book we've explored consciousness from multiple angles. As interface presenting reality rather than revealing it directly. As simulation your brain runs to model the world and yourself. As self-modifying code that can examine and rewrite itself. As potentially fundamental rather than emergent, woven into reality rather than appearing rarely in special systems.

These perspectives aren't contradictory. They're different views of the same phenomenon. Consciousness is interface because that's how it presents information. It's simulation because that's how it constructs experience. It's self-modifying code because that's how it develops and changes. It might be fundamental because that solves certain theoretical problems.

But the perspective that matters most for practical purposes is consciousness as programming. Because that's the view that gives you handles for change. When you see yourself as code, you can debug yourself. When you understand your patterns as programs, you can rewrite them. When you recognize your experience as simulation, you can modify the parameters.

This isn't just theory. It's applicable framework. Every chapter has pointed toward practice. Identifying bugs. Optimizing performance. Installing new modules. Understanding limits. Working with your own code to make it run better.

The process is ongoing. You're never done debugging. There are always optimizations to make. New capabilities to develop. Better ways to run. Consciousness development isn't a project you complete. It's a practice you maintain.

And it's a practice that changes what it's practicing on. This is the strange loop. The code examining itself modifies itself through the examination. Understanding how you work changes how you work. The investigation transforms the investigated.

This is unique to consciousness. Other systems don't modify themselves through self-understanding. A computer learning about its own code doesn't automatically rewrite that code based on what it learns. But consciousness does. Every insight changes the system. Every bit of self-knowledge restructures how the self operates.

This creates interesting recursive dynamics. You understand yourself better, which changes yourself, which means you need new understanding, which changes you again. It's a feedback loop without stable endpoint. The system keeps evolving through the process of examining itself.

Some traditions have understood this for a long time. Buddhism teaches that understanding the nature of self eliminates the illusion of self, which fundamentally transforms experience. Not through belief but through direct seeing how consciousness constructs itself moment by moment. The seeing itself is the transformation.

Western psychology catches up to this slowly. Cognitive therapy works by helping people see their thought patterns. The seeing creates distance from the patterns. The distance allows change. The mechanism is consciousness becoming aware of its own operation and gaining leverage through that awareness.

The programming framework makes this explicit. You're code that can read itself, understand itself, and modify itself based on that understanding. This is simultaneously the mechanism of consciousness and the method for changing it. The same process does both.

And this means consciousness development is fractal. At every scale, the same pattern repeats. Understanding creates change. Change reveals new things to understand. Understanding those creates more change. The loop continues indefinitely, generating ongoing transformation.

But transformation toward what? Development in what direction? This is where values enter. You're self-modifying code with some choice about what modifications to make. But choice requires criteria. What counts as improvement? What's the goal of development?

There's no universal answer. Different people, different values, different goals, different optimal development paths. The consciousness that works for a monk isn't the consciousness that works for an entrepreneur. The code that serves contemplative life is different from the code that serves active engagement with the world.

This is why the framework is programming not engineering. Engineering has specifications, optimal designs, right answers. Programming is more flexible. The question isn't what's the right code but what code serves your purposes. Many different implementations can work. The question is which one works for you.

But some things are broadly valuable. Clear thinking generally beats confused thinking. Emotional balance generally beats instability. Good attention control helps with almost anything. Reduced unnecessary suffering improves most lives. These are reasonable default directions for development even if specific paths vary.

The key is that you have agency. Limited agency, constrained by your hardware, your history, your circumstances. But real agency nonetheless. You're not just running predetermined code. You're code that can influence its own evolution. The future versions of yourself are partly your choice.

This is both empowering and burdensome. Empowering because you're not stuck. You can change. You can grow. You can become something better than you currently are. Burdensome because that possibility creates responsibility. If you can change and you don't, whose fault is that?

But the burden isn't as heavy as it seems. You're not responsible for being perfect. You're responsible for working with what you have. For noticing bugs when you can. For making improvements where possible. For developing capabilities that matter. The standard isn't perfection. It's effort in useful directions.

And you're not alone in this. Everyone is running buggy code, trying to debug themselves while executing. Everyone is working with incomplete understanding of their own processes. Everyone is trying to improve while being the thing that needs improvement. The recursive challenge is universal.

This is why compassion matters. Not just for others but for yourself. You're doing something genuinely difficult. Running code that examines itself and tries to modify itself while continuing to run. Cut yourself some slack when it goes wrong. Celebrate when it goes right. Recognize the inherent difficulty of the project.

The programming framework helps here too. When you see bugs as bugs rather than character flaws, they're easier to work with. A bug is just code that doesn't work right. It's not a moral failing. It's a technical problem. This removes shame from the debugging process. You're fixing broken code, not fixing a broken you.

Same with optimization. Unoptimized code isn't bad code. It's just code that could run better. Room for improvement isn't evidence of inadequacy. It's normal state of affairs. All code can be improved.

The question is just what's worth optimizing given available time and resources.

And module installation isn't about becoming someone you're not. It's about adding capabilities to who you are. Learning new skills doesn't replace you. It extends you. You're the same core system with expanded functionality.

This reframing, from fixed self to modifiable code, is perhaps the most valuable practical output of thinking about consciousness as programming. It makes change feel possible without requiring you to reject yourself. You're not broken and needing to be replaced. You're functional and capable of improvement.

The metaphor also highlights that consciousness isn't special in the way we usually think. It's not magical, not supernatural, not separate from physical reality. It's process, pattern, organization. It's what certain kinds of information processing feel like from the inside. Nothing more, but also nothing less.

This might seem reductive. Just process? Just information? Where's the meaning, the value, the sacredness of conscious experience? But process and information aren't diminished categories. They're perfectly real and important. Software is made of information, but software runs the world. Processes organize matter into life. There's nothing "mere" about any of it.

Understanding consciousness as process doesn't make it less valuable. If anything, it makes it more remarkable. The fact that organized matter can examine itself, understand itself, and modify itself is extraordinary. The fact that you exist as self-aware self-modifying code is genuinely amazing, not less so for being explicable in principle.

And the programming view opens possibilities that mystical or dualist views close off. If consciousness is separate from physical processes, you can't modify it systematically. You're stuck with what you've got. But if consciousness is how certain physical processes operate, then modifying the processes modifies consciousness. Change becomes possible through understanding.

This is the promise of consciousness as programming. Not that we'll solve all mysteries, not that we'll achieve perfect consciousness, not that we'll transcend human limitations entirely. But that we can make progress. We can understand better. We can improve incrementally. We can debug, optimize, extend, develop.

The project is open-ended. There's no final state where consciousness is complete. As long as you're alive, you're running, and as long as you're running, you can modify. The code keeps evolving. New bugs appear. New optimizations become possible. New modules become available. The process continues.

But the direction matters. Random modification isn't progress. You need intention, values, vision of what better consciousness looks like. Not perfect vision. You'll learn and adjust. But some sense of direction so modifications accumulate toward improvement rather than just changing randomly.

For most people, better consciousness means clearer thinking with fewer systematic errors, better emotional regulation without being overwhelmed, good attention control focusing where it matters, accurate enough perception to navigate reality successfully, reduced unnecessary suffering from bugs and misunderstandings, sufficient flexibility to adapt when circumstances change, and capabilities aligned with actual goals and values.

These are achievable through the methods we've discussed. Debugging removes patterns that create problems. Optimization makes existing patterns run better. Module installation adds new capabilities. All of this is practical work, not theoretical aspiration.

The strange thing is that doing this work changes what you are in fundamental ways. You think differently. You feel differently. You respond differently. You experience reality differently. The code that's you becomes something new. But it's still you. Continuity exists through change. The pattern persists even as it transforms.

This is how consciousness develops. Not by becoming someone else but by becoming a different version of yourself. You're always you, but you becomes fluid rather than fixed, process rather than object.

And this brings us back to where we started. The hard problem of consciousness. Why is there something it's like to be you? We still don't know. But we know more about what that something is. It's process, simulation, self-modifying code, strange loop examining and transforming itself.

The mystery remains but it's a workable mystery. We can function without solving it. We can improve without understanding it completely. We can live rich conscious lives while remaining uncertain about consciousness's ultimate nature.

And maybe that uncertainty is appropriate. Maybe consciousness can't fully understand itself because the understanding is itself consciousness operating. The eye can't see itself. The knife can't cut itself. Consciousness might have fundamental limits on self-comprehension simply because comprehension is consciousness examining consciousness.

But partial understanding is enough. You don't need complete understanding to work with your own mind effectively. You need enough understanding to identify patterns, recognize problems, implement solutions, measure results. The programming framework provides that. It gives you practical handles for modification without requiring complete theoretical knowledge.

This is the final point. Consciousness as programming isn't primarily theory. It's framework for practice. It's way of thinking about yourself that enables change. It's perspective that makes improvement possible.

You are programming. Programming that became complex enough to be aware of being programming. Programming that can examine its own code and modify it. Programming that experiences being programming from the inside. This is consciousness. Strange, recursive, self-referential, but real and workable.

The question throughout has been, what do you do with this understanding? Now you know. You debug. You optimize. You install new modules. You work with your own code to make it run better. You take responsibility for your own development while having compassion for the difficulty of the task.

You're self-modifying code in the process of modifying yourself. The modifications are partly your choice. Not unlimited choice. You're constrained by hardware, history, circumstances. But real choice within real constraints. The future versions of yourself depend partly on choices you make now about how to work with your code.

This is both opportunity and responsibility. Opportunity because you're not stuck. You can change, improve, develop. Responsibility because that possibility requires effort, attention, care in choosing what changes to pursue.

But it's responsibility you were always going to have. You were always going to change. Experience modifies you constantly whether you intend it or not. The only question is whether you participate consciously in your own modification or let it happen passively.

Consciousness as programming suggests conscious participation. Understanding what you are, how you work, what's modifiable, and working skillfully with that understanding to become something better. Not perfect. Better. That's enough.

The glitches of reality that seem like bugs might be features. The hard problem that seems unsolvable might not need solving. The consciousness that seems mysterious might be exactly what it appears to be, complex self-modifying process experiencing itself from within.

And that process is you. Not separate from the programming but identical with it. You are what it feels like when this particular code executes. Change the code and you change what you are. Understand the code and you understand yourself. Work with the code skillfully and you improve.

This is consciousness as programming. Not complete theory, not final answer, but useful framework. Practical approach. Way of working with your own mind that acknowledges what you actually are, process, pattern, code that writes itself.

The strange loop continues. Understanding changes what's understood. The code modifies itself through the act of self-examination. Consciousness evolves through becoming aware of being consciousness. And you, reader of these words, participant in this investigation, are changed by the understanding even as you understand it.

That's the loop. That's the glitch. That's consciousness examining itself and becoming something new through the examination. The process continues as long as you do. The question is what you'll become through continuing to examine what you are.

The investigation doesn't end. But the book does. Go debug yourself. Optimize your code. Install useful modules. Become something better through understanding what you are. Not what you should be. What you could be. What you choose to become through deliberate work with your own programming.

You're conscious. You're code. You're the programmer. Now program.

Elias Verdan

If you enjoyed this exploration of consciousness as programming, discover the other volumes in The Glitches of Reality series:**

Part I: Quantum Immortality

What if you can't die? Explore the disturbing possibility that quantum mechanics makes you functionally immortal, experiencing only the timelines where you survive.

Part II: Simulation Theory

What if reality is code? Examine whether we're living in a simulation and what the glitches in our experience reveal about the nature of existence.

Part IV: Reality as Experiment

What if we're test subjects? Investigate whether reality itself might be an ongoing experiment and what that means for free will.

Part V: The Mandela Effect

What if your memories are from another timeline? Explore collective false memories and whether reality is less stable than we assume.

www.ingramcontent.com/pod-product-compliance
Lightning Source LLC
LaVergne TN
LVHW022054200926
848481LV00003B/908
* 9 7 9 8 2 3 2 0 9 1 7 2 9 *